Alexander Winkler

Transient Behaviour of ITER Poloidal Field Coils

**Karlsruher Schriftenreihe zur Supraleitung
Band 002**

Herausgeber: Prof. Dr.-Ing. M. Noe
                   Prof. Dr. rer. nat. M. Siegel

Transient Behaviour of ITER Poloidal Field Coils

by
Alexander Winkler

Dissertation, Karlsruher Institut für Technologie
Fakultät für Elektrotechnik und Informationstechnik, 2010
Hauptreferenten: Prof. Dr.-Ing. Mathias Noe
Korreferent: Prof. Dr.-Ing. Thomas Leibfried

Impressum

Karlsruher Institut für Technologie (KIT)
KIT Scientific Publishing
Straße am Forum 2
D-76131 Karlsruhe
www.ksp.kit.edu

KIT – Universität des Landes Baden-Württemberg und nationales
Forschungszentrum in der Helmholtz-Gemeinschaft

KIT Scientific Publishing 2010
Print on Demand

ISSN 1869-1765
ISBN 978-3-86644-595-6

Transient Behaviour of ITER Poloidal Field Coils

Zur Erlangung des akademischen Grades eines
DOKTOR-INGENIEURS
von der Fakultät für
Elektrotechnik und Informationstechnik
der Universität Karlsruhe (TH)
genehmigte
DISSERTATION
von
Dipl.-Ing. Alexander Winkler
geb. in: Krasnoturinsk (Russland)

Tag der mündlichen Prüfung: 2. November 2010
Hauptreferent: Prof. Dr.-Ing. Mathias Noe
Korreferent: Prof. Dr.-Ing. Thomas Leibfried

Danksagung

Die vorliegende Arbeit entstand während meiner Tätigkeit als Doktorand am Institut für Technische Physik (ITEP) des Karlsruher Instituts für Technologie (KIT). Mein besonderer Dank geht an Herrn Prof. Dr.-Ing. Mathias Noe für die Übernahme des Hauptreferats, seine fachliche Betreuung und motivierende Unterstützung meiner Arbeit.

Für das Korreferat dieser Arbeit bedanke ich mich beim Herrn Prof. Dr.-Ing. Thomas Leibfried vom Institut für Elektroenergiesysteme und Hochspannungstechnik (IEH) des Karlsruher Instituts für Technologie (KIT).

Dem Leiter des Hochspannungslabors des ITEP, Herrn Stefan Fink, möchte ich für seine wissenschaftliche Betreuung, die allgemeine und fachliche Diskussionen besonders danken.

Mein herzlicher Dank gilt auch den Mitarbeitern der Gruppen „Fusionsmagnete" und „Höchstfeldmagnete" und insbesondere den Herren Volker Zwecker, Patrick Heinrich und Uwe Braun für Ihre tatkräftige Unterstützung beim durchführen der Laborversuche. Den Gruppenleitern Herrn Dr. Walter H. Fietz und Herrn Dr. Theo Schneider möchte ich insbesondere für Ihr persönliches Engagement und fachliche Beratung danken. Genauso möchte ich mich für die anregenden Diskussionen bei allgemeinen und wissenschaftlichen Fragestellungen bei Herren Dr. Klaus-Peter Weiss, Dr. Michael Schwarz, Dr. Christian Schacherer, André Berger und Olaf Mäder bedanken.

Schließlich, möchte ich allen Menschen aus meinem persönlichen Umfeld für ihre Hilfe meinen Dank aussprechen und einen besonderen Dank an meine Familie für die entgegengebrachte Geduld und Unterstützung meiner Arbeit richten.

Karlsruhe,
September 2010

Alexander Winkler

Kurzfassung

Der International Thermonuclear Experimantal Reactor (ITER) ist ein internationales Projekt, der die großtechnische Nutzung der Kernfusion vorbereiten soll. Das supraleitende Spulensystem von ITER wird zum Einschluss und zur Steuerung des Plasmas im Fusionsreaktor eingesetzt. Transiente elektrische Spannungen entstehen an den Anschlüssen der supraleitenden Spulen und können zu internen Schwingungen und Spannungsüberhöhungen führen. Die Auslegung der Hochspannungsisolierung ist notwendig für einen zuverlässigen Betrieb der Spulen bei verschiedenen Betriebsszenarien und möglichen Fehlerfällen. Ziel dieser Arbeit war die Berechung des transienten elektrischen Verhaltens und der internen Spannungsverteilung der ITER Poloidal Feld (PF) Spulen bei Nennbetrieb, der Schnellentladung und definierten Fehlerfällen und wurde am Beispiel der PF 3 und PF 6 Spulen durchgeführt. Mit den berechneten internen Maximalspannungen werden zukünftig die Amplituden und Spannungsformen für die Tests der Hochspannungsisolierung festgelegt.

Um einen ersten Eindruck über das transiente elektrische Verhalten zu bekommen, wurden zuerst die Resonanzfrequenzen der beiden Spulen im Frequenzbereich berechnet. Dabei wurde der Einfluss der symmetrischen und unsymmetrischen Erdung und der Instrumentierungsleitungen auf die Resonanzfrequenz untersucht.

Die Berechungen der Resonanzfrequenz der Spulen und die interne Spannungsverteilung wurden mit einer speziellen Berechnungsstrategie durchgeführt. Dabei wurden die Frequenzabhängigkeit der Induktivitäten und die Spannungsverläufe an den Spulenanschlüssen berücksichtigt. Diese Spannungsverläufe bestehen aus mehreren überlagerten Spannungen mit unterschiedlichen Frequenzen. Die Finite Elemente Methode (FEM) und Netzwerkmodelle der Spulen konnten mit den entsprechenden Programmen nur für diskrete Frequenzen aufgestellt werden, somit wurden die Berechnungen in mehreren Teilbereiche aufgeteilt.

Die frequenzabhängige Induktivitäten der Spulen wurden mit einem FEM Programm für relevante Frequenzen berechnet und in das detaillierte Netzwerkmodell der Spulen als konzentrierte Elemente eingesetzt. Die aufgestellten Netzwerkmodelle wurden sowohl für die Berechnung der Resonanzfrequenz als auch der internen Spannungsverteilung benutzt. Um den Einfluss der Frequenzabhängigkeit der Induktivitäten auf die interne Spannungsverteilung der beiden Spulen zu untersuchen, wurden die Netzwerkmodelle für die Resonanzfrequenz und zusätzlich für eine vergleichsweise niedrige bzw. eine hohe Frequenz aufgestellt.

Die interne Spannungsverteilung wurde für folgende vier Szenarien im Zeitbereich berechnet: das Referenzszenario, die Schnellentladung und zwei Fehlerfälle. Hierbei stellt das Referenzszenario den Nennbetrieb der Spulen dar. Bei der Schnellentladung werden alle PF Spulen gleichzeitig an ihren jeweiligen Entladewiderständen entladen. Bei den zwei Fehlerfällen wurde bei jeweils unterschiedlichen Strom- und Spannungswerten ein Erdschluss am negativen Anschluss der PF 3 bzw. PF 6 Spule während der Schnellentladung angenommen.

Die maximale Spannungen an den Anschlüssen und auch die internen Spannungen, die an der Erd-, Lagen- und Windungsisolation der PF 3 und PF 6 Spulen berechnet wurden, sind höher, als die bisher angegebenen Spannungswerte für die ITER PF Spulen [Lib08, DDD06h].

Bei schnellen transienten Spannungsanregungen an den Spulenanschlüssen, die während der Schnellabschaltung und der beiden Fehlerfälle auftreten, wurde nichtlineare Spannungsverteilung innerhalb der Spule berechnet. Dadurch waren die berechneten Spannungen an der Erdisolation der PF 3 und PF 6 Spulen teilweise höher, als die Spannungen an den Spulenanschlüssen.

Table of Contents

1   Motivation and Introduction

The supply of safe and sustainable electrical energy is one of the most important issues for the mankind today and in the future. Fusion could take a big part in the centralised base load power generation of electrical energy. The International Thermonuclear Experimental Reactor (ITER) is the first attempt to operate a fusion reactor with fusion power in amounts comparable with that of today's conventional electrical power plants. One of the main parts of ITER is the superconducting coil system which will be used for confinement and control of plasma during the fusion reaction. The verification of the high-voltage insulation co-ordination of the coil system is essential for a reliable operation of ITER. Transient electrical excitations occur on the terminals of superconducting coils, for example, in case of a fast discharge of the coils and if a failure appears in the components of the electrical circuit. Transient voltage on the coil terminals may lead to non-linear voltage distribution and oscillations within large superconducting coils because of their large dimensions and high number of turns. Internal voltages may even be higher than the voltages on the coil terminals. This effect depends on the amplitude and rise time of the excitations and was measured within the ITER Toroidal Field Model Coil [Fin02] which is about three times smaller than the ITER toroidal field (TF) coil. The calculations of the transient electrical behaviour of the coils provide a basis for high-voltage insulation co-ordination and definition of test voltages and waveforms. The high-voltage tests during the manufacturing process will control the quality of the electrical insulation and ensure reliable operation of the coils.

The main objective of this work was to calculate the transient electrical behaviour of ITER poloidal field (PF) coils using the PF 3 and PF 6 coils as examples. The importance of the calculations is underlined by the fact that the replacement of one PF coil will take several years, if a fault occurs in the insulation of the coil and continuous operation will not be possible. Internal voltage distribution was analysed for the PF 3 coil, because this PF coil has the largest coil diameter and for the PF 6 coil, because it is the PF coil with the highest number of turns.

The inductances of the PF coils are frequency-depended and decrease with increasing frequency. Different Finite Element Method (FEM) and network models for discrete frequencies were used to calculate the inductances of each turn and the internal voltage distribution for excitations at different frequencies. The voltages on the terminals of superconducting coils can usually be neglected during operation with constant current under steady-state conditions. Transient voltage waveforms occur on coil terminals, if the change of the current is high enough. The fall time of the coil current during fast discharge is in the range of seconds. Due to the commutation of coil current to a discharge resistor, the calculated voltage excitations on the coil terminals had high voltage amplitudes and rise times in the range of microseconds. These voltage excitations cause oscillations and reflections which consist of numerous frequencies in several frequency ranges up to a few hundred kHz. The main challenge in the calculations was to deal with relatively slow current alternations in the range of seconds, the relatively fast voltage excitations on the coil terminals in the range of microseconds and the frequency dependence of the inductances of the PF coils. Thus, a special calculation strategy was defined for calculation of internal transient voltage distribution which is described in detail in chapter 3. The frequency-depended inductances of the coils were calculated with FEM models. The current behaviour and the voltage waveforms on coil terminals were calculated with DC network models of the power supply circuits of the coils. The internal voltage distribution was calculated with detailed frequency depended network models of the PF 3 and PF 6 coils for low, resonance and high frequencies.

In the following chapters, the calculation process is described in detail. The overview of the ITER coil system is given in chapter 2. The detailed and simplified FEM models of the coils are described in chapter 4. In chapter 5, the network model of the CS PF coil system and detailed network models of the PF 3 and PF 6 coils are shown in detail. The results of the calculations in the frequency and time domain are discussed in chapter 6. The detailed data of the FEM calculations and network calculations are summarised in the annex.

2 Overview of the ITER Coil System

ITER will be built for the technical and scientific verification of the feasibility of fusion power plants. One of the main parts of ITER is the superconducting coil system. For a detailed analysis of the internal voltage distribution within the coils, detailed information on the ITER coil system is necessary, which is given in this chapter. The first section gives a short overview of the tokamak coil system. The following sections describe the dimensions of the CS PF coil system, the internal design of the PF coils and the materials which will be used for the coil construction in detail. The electrical circuits for coil power supply are described in the last section. Further information about ITER and fusion can be taken from [ITER01].

2.1 The Tokamak Coil System

ITER is based on the tokamak coil system. Tokamak is Russian and means toroidal field with magnetic coils. The ITER tokamak coil system consists mainly of three different superconducting coil systems. The scheme of the ITER coil system is shown in Fig. 2.1. The 18 Toroidal Field (TF) coils are D-shaped coils for confinement of the plasma inside the vacuum vessel. The six Poloidal Field (PF) coils will control the position and shape of the plasma in the vacuum vessel. The Central Solenoid (CS) coil consisting of six identical CS coil modules will induce current in the plasma. Both the CS and PF coils will be built with rotational symmetry.

The coil system has an outer diameter of nearly 25 m and a total height of more than 16 m. All TF coils will be built identically and have a total height of more than 13 m. The TF coils will be based on an Nb_3Sn superconductor. Each coil has 134 turns and will be driven with operation current of 68 kA and a nominal magnetic peak field of nearly 12 T. The outer case of TF coils will be used for vertical loads of the PF coils. The vertical load support for the CS coil is inside the torus which will be shaped by 18 TF coils [DDD06].

The CS coil consists of 6 identical coil modules which are stacked on top of each other to form one coil. The two coil modules in the centre are switched in one serial power supply circuit. The other coil modules will be driven by own power supply circuits. Each CS coil module has 549 turns and will be built with an Nb_3Sn superconductor. The mean diameter of one coil module is 3.5 m and the height is 2.1 m. The operating current of the CS coil is 45 kA and the nominal magnetic peak field is 13 T.

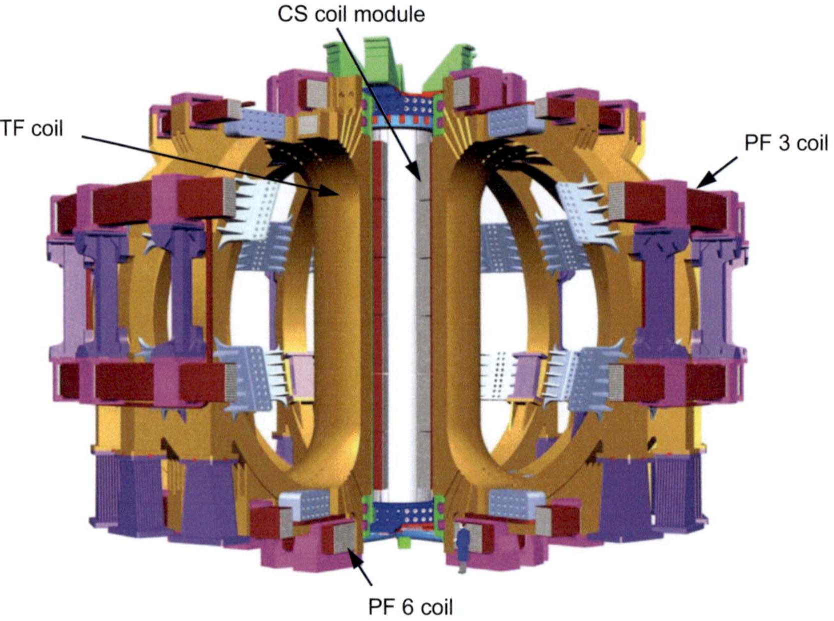

Fig. 2.1: ITER coil system consisting of 18 Toroidal Field (TF) coils, 6 Poloidal Field (PF) coils and 6 Central Solenoid (CS) coil modules [ITER].

All PF coils have different dimensions and numbers of turns. The PF 1 coil has the smallest mean diameter of 8 m, the PF 3 coil has the largest mean diameter of 24 m. The number of turns varies from 106 in the PF 2 to 425 in the PF 6 coil. The operating current of each PF coil is 45 kA, despite the PF 2 coil with 41 kA, and the nominal magnetic peak field is up to 6 T. The coils will be made of an NbTi super-conductor.

The design of the CS and PF coil system, the PF conductor configuration and the power supply circuits of CS and PF coils are described in detail in the following sections.

2.2 Design of Central Solenoid and Poloidal Field Coils

The design and location of the Central Solenoid (CS) and Poloidal Field (PF) coils was based on the requirements of plasma physics. The six PF coils will control the position and shape of the plasma in the vacuum vessel. The CS coil consisting of six CS coil modules will induce current in the plasma. The TF coils were neglected in the calculations of the transient behaviour of the PF coils, because the magnetic field caused by TF coils is orthogonal to the magnetic fields of the CS and PF coils.

The rotational symmetry of the CS and PF coils was used to build two-dimensional models of the CS PF coil system. Fig. 2.2 shows the cross-section of the CS PF coil system with the rotational symmetry axis and the magnet centre line. The magnet centre line is defined to be located between the upper and the lower CS coil modules. The dimensions of the coils in Z-direction are specified in relation to the magnet centre line [DDD06]. Fig. 2.3 shows the cross-section of the PF 6 coil with the layer and turn numbering scheme.

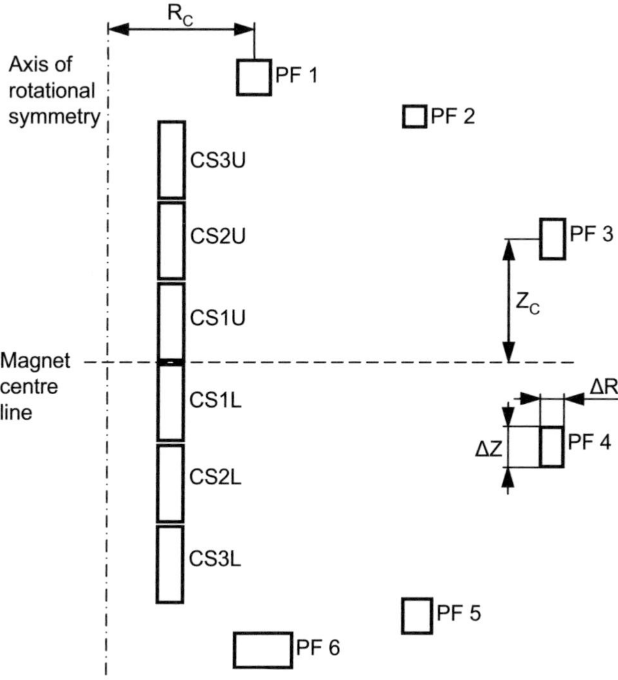

Fig. 2.2: Cross-section of the ITER CS PF coil system with designations R_C, Z_C, ΔZ and ΔR used for coil dimensions in Tab. 2.1.

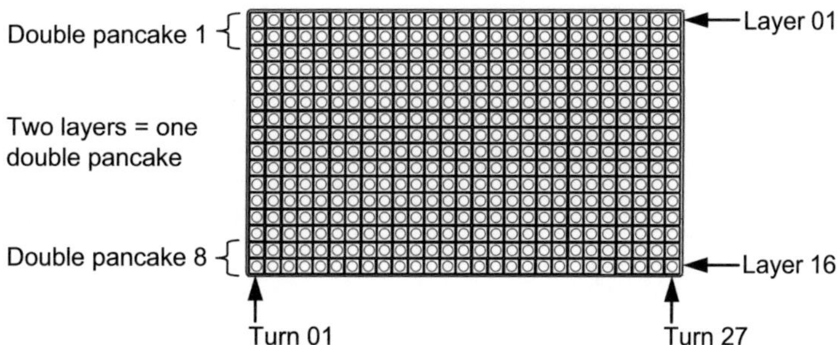

Fig. 2.3: Cross-section of the PF 6 coil with the layer and turn numbering scheme.

Each PF coil, with the exception of PF 2, consists of 16 layers. Two layers make up a so-called double pancake. Thus, each coil has 8 double pancakes with the exception of PF 2 coil which consists of 5 double pancakes only. Each CS coil module consists of 6 hexa pancakes and one quad pancake.

The CS coil has a conducting, hard grounded surface at the outside of each CS coil module [DDD06j]. The PF coils have protection covers made of stainless steel [PPA3]. The magnetic coupling between the coils at higher frequencies will be weakened because of the metallic coil cases and other metallic parts between the coils. The eddy current which will be induced in these metallic parts reduces the coupling to negligible values at higher frequencies. At frequencies higher than 5 kHz, every coil was considered as a coil without magnetic coupling to other coils, which is explained in detail in section 4.2. The designation of the CS coil modules was given beginning from the magnet centre line to the top and bottom from 1 to 3. Upper modules are abbreviated by an U behind coil names, lower modules by an L, e. g. CS3U is the uppermost CS coil module. The PF coils are numbered from top to bottom. The dimensions of CS and PF coils at 4.5 K are summarised in Tab. 2.1. The positions of the coil centres in Z-direction are specified from the magnet centre line of the coil system, which is shown in Fig. 2.2.

Tab. 2.1: Location and size of PF coils and CS coil modules at 4.5 K [DDD06a].

Coil number	Position of coil centre		Coil size without ground insulation and protector wrap	
	R_C	Z_C	ΔR	ΔZ
PF 1	3.943 m	7.557 m	0.968 m	0.976 m
PF 2	8.319 m	6.530 m	0.649 m	0.595 m
PF 3	11.997 m	3.265 m	0.708 m	0.966 m*
PF 4	11.967 m	-2.243 m	0.649 m	0.966 m*
PF 5	8.395 m	-6.730 m	0.820 m	0.945 m
PF 6	4.263 m	-7.557 m	1.633 m	0.976 m
CS3U	1.772 m	5.355 m	0.719 m	2.092 m
CS2U	1.772 m	3.213 m	0.719 m	2.092 m
CS1U	1.772 m	1.071 m	0.719 m	2.092 m
CS1L	1.772 m	-1.071 m	0.719 m	2.092 m
CS2L	1.772 m	-3.213 m	0.719 m	2.092 m
CS3L	1.772 m	-5.355 m	0.719 m	2.092 m

* For PF 3 and PF 4 coils the dimensions in Z-direction in [DDD06a] are specified for the design with separator plates which will not be used in the current design. The Z-dimensions of PF 3 and PF 4 coils were calculated with dimensions specified in [DDDD2] for room temperature and scaled down with a shrinkage factor of -0.29 % at 4.5 K [DDD06g].

2.3 Configuration of Poloidal Field Coils

The design of all PF coils is described in this section, although the PF 3 and PF 6 coils only will be analysed in detail in the calculations of the transient behaviour of ITER PF coils. All PF coils will be built with an NbTi superconductor which is cooled by liquid helium. The PF coils are designed for a normal operation current of 45 kA at an operating temperature of 5 K. In the backup mode, when one of the damaged double pancakes of a coil will be disconnected and by-passed, the coil current will be increased to 52 kA. The operating current of the PF 2 coil under normal conditions is 41 kA, although the conductor is designed for 45 kA. The PF 2 coil has 5 double pancakes only. In the backup mode, with one double pancake less, the coil current will exceed the maximum current limited to 52 kA. Therefore, normal operating current has to be limited to 41 kA. The detailed data of the PF coil conductors are summarised in Tab. 2.2.

Tab. 2.2: Detailed data of PF conductors [DDD06b].

	PF 1 & PF 6	PF 2, PF 3 & PF 4	PF 5
Coolant normal / backup	Inlet 4.7 K/4.4 K	Inlet 4.7 K	Inlet 4.7 K
Type of strand	NbTi	NbTi	NbTi
Operating current (kA) normal / backup	45 / 52	45* / 52	45 / 52
Nominal peak field (T) normal / backup	6.0 / 6.4	4.0	5.0
Operating temperature (K) normal / backup	5.0 / 4.7	5.0	5.0
I_{OP}/I_C (ratio of operating current to critical current) normal / backup	0.127 / 0.144	0.365 / 0.422	0.264 / 0.305
Cable diameter (mm)	38.2	34.5	35.4
Central spiral outer x inner diameter (mm)	12 x 10	12 x 10	12 x 10
Conductor outer dimensions (mm)	53.8 x 53.8	52.3 x 52.3	51.9 x 51.9
Jacket material	Stainless steel 316L	Stainless steel 316L	Stainless steel 316L
Superconducting strand diameter (mm)	0.73	0.73	0.72
Superconducting strand Cu:non-Cu ratio	1.6	6.9	4.4
Cabling pattern (+ Cu core)	3x4x4x5x6	((3x3x4+1)x4+1) x6	((3x3x4+1)x5+1) x6
Number of superconducting strands	1440	864	1080
Local void fraction (%) on strand bundle	34.5	34.2	34.3
Superconducting strand weight conductor (kg/m)	4.885	2.931	3.564

* For the PF 2 coil the operation current under normal conditions is 41 kA

The conductor in the PF coils is cooled by helium which flows inside the cooling tube in the centre of each conductor. The cooling tube has an inner diameter of 10 mm and an outer diameter of 12 mm and is designed as a spiral to facilitate helium flow into the superconducting cable. The superconducting cable is placed directly on the outside of the cooling tube. The cross-section and relevant dimensions of the PF conductors are shown in Fig. 2.4. Fig. 2.5 shows a conductor which will be used in ITER PF coils. The superconducting cable consists of up to 1440 superconducting strands [DDD06b], which are separated into five stages of sub-bundles, the cabling pattern for the superconducting cable of each PF coil conductor is given in Tab. 2.2. The sub-bundles in each stage are twisted around each other to make the current density in the superconducting strands and sub-bundles more homogeneous and prevent individual local overload of one strand or one bundle.

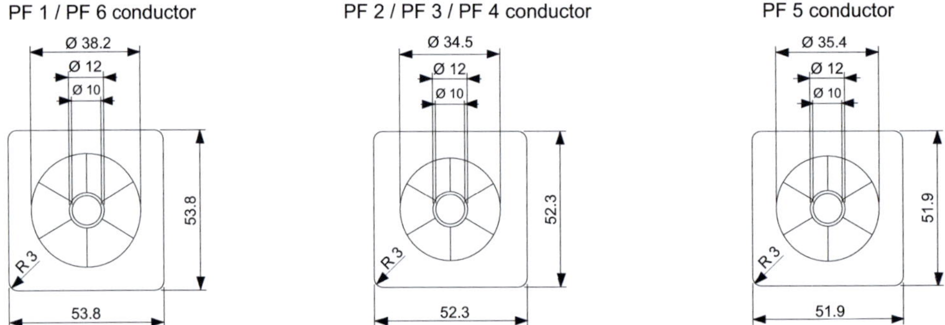

Fig. 2.4:   Dimensions of the PF coil conductors for PF 1 to PF 6 coils [DDDD1].

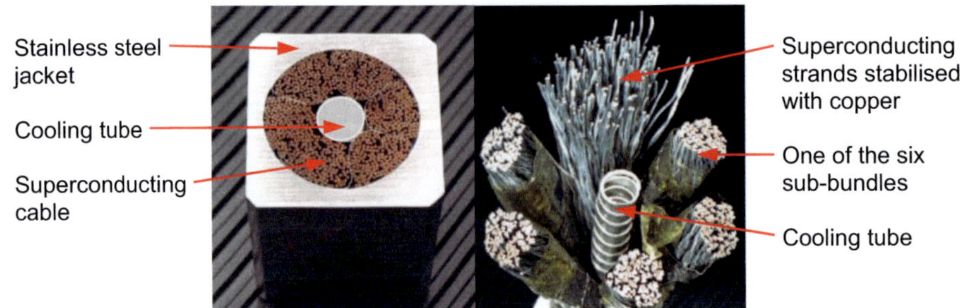

Fig. 2.5:   Conductor which will be used in ITER PF coils [ITER].

Copper strands in superconducting cables are necessary to have an additional current path in case of a quench to stabilise the superconductor. The superconducting strands have different copper: non-copper ratios, depending on the PF coil in which they are used. Different sub-bundles of the superconducting cable have additional copper strands to enlarge the copper ratio of the cable, shown in Tab. 2.2. For mechanical stabilisation the superconducting cables are pulled into a square stainless steel jacket. For protection during the pull-through into the jacket, the superconducting cable is covered with 0.08 mm stainless steel wrap [DDD06c]. The outer dimensions of the jacket are between 51.9 mm and 53.8 mm depending on the PF coil and shown in Fig. 2.4.

The insulation of the PF coils consists of several insulation types, the conductor, layer and ground insulation. Fig. 2.6 shows the dimensions of the insulation of the PF 6 coil, the dimensions of insulations of all PF coils are summarised in Tab. 2.3. The turn insulation consists of two insulation layers of 1.5 mm each [DDD06d]. It is made up of half overlapped interleaved polyimide film and dry glass. After the winding of the double pancakes, the turn insulation will be vacuum-impregnated by epoxy resin. After the impregnation, the adjustable bonding layer is placed between double pancakes. Five double pancakes for the PF 2 coil and 8 double pancakes for other PF coils are stacked to one winding pack. The ground insulation with a compacted thickness of 8 mm is wrapped around the whole winding pack [DDD06d]. The ground insulation consists of 0.25 mm thick glass, 0.05 mm thick polyimide film and 0.25 mm thick glass wrapped in nine 50 % overlapped layers [DDD06d]. The total uncompacted thickness of the ground insulation is 9.9 mm. In the second impregnation step, the ground insulation is also vacuum-impregnated with epoxy resin. In addition to the ground insulation, a protective cover of 2 mm thickness made of pre-cured G10 glass epoxy composite encases the PF coils [DDD06d].

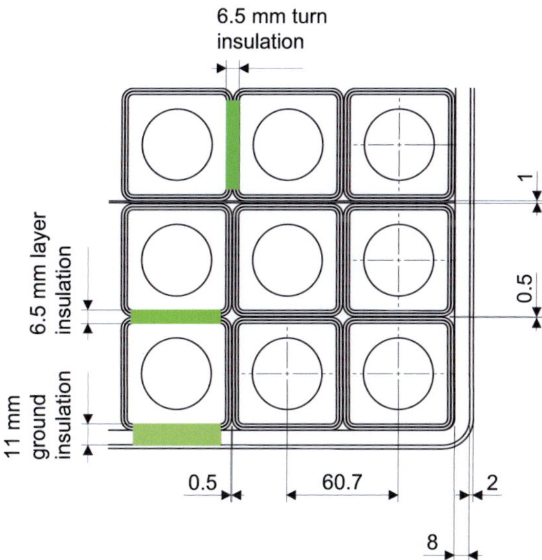

Fig. 2.6:   Conductor and insulation dimensions of the PF 6 coil [DDDD2].

Tab. 2.3:   Insulation dimensions of PF coils [DDDD2].

	PF 1 / PF 6	PF 2	PF 3 / PF 4	PF 5
Turn insulation	3.0 mm	3.0 mm	3.0 mm	3.0 mm
Adjustable turn insulation	0.5 mm	0.5 mm	0.5 mm	0.5 mm
Cumulative turn-to-turn insulation	6.5 mm	6.5 mm	6.5 mm	6.5 mm
Distance between the turn centres	60.7 mm	59.2 mm	58.8 mm	59.2 mm
Adjustable layer insulation	0.5 mm	0.5 mm	1.0 mm	0.5 mm
Adjustable bonding layer	1.0 mm	1.0 mm	2.0 mm	1.0 mm
Ground insulation	8.0 mm	8.0 mm	8.0 mm	8.0 mm
Protection wrap	2.0 mm	2.0 mm	2.0 mm	2.0 mm

The PF coils are designed as double pancake winding with the two-in-hand winding configuration. The two-in-hand winding scheme and stacking of double pancakes allows for all joints to be located on the outer pancake diameter, as shown in Fig. 2.7. The inlets of liquid helium are on the inner side of the double pancakes, as the consequence the inner turn of the coils is supplied always with colder coolant than the outer turn because the magnetic field of PF coils is higher on the inner turn of the coils. The two-in-hand winding also allows the high-voltage test of turn insulation of a double pancake before the connection of the internal double pancake joint, e.g. the first conductor will be connected to high voltage and the second conductor to ground. An additional advantage of this winding configuration is that the possibility of a short circuit between the turns of the same conductor within the same pancake is strongly reduced, because they are separated by the second conductor, which simplifies the search for damaged turn to turn insulation [CDA4].

The conductors 1 and 2 will be first wound inwards in the upper pancake. If the upper pancake is wound completely, both conductors will be transposed to lower pancake and their position will be changed. Then, they will be wound outwards till the lower pancake is completed. The end of conductor 1 in the lower pancake will be connected by the internal pancake joint with conductor 2 in the upper pancake. The ends of conductor 1 in the upper pancake and conductor 2 in the lower pancake will be connected by external double pancake joints to the next upper and lower double pancakes, respectively.

The joints between the two layers within a double pancake (internal pancake joint) as well as joints between neighbouring double pancakes (external double pancake joints) are placed outside of the winding pack. Having all joints on the outer diameter of the coils, it is possible to bypass a double pancake in case of a short fault by opening three joints and reconnecting two of the joints. The bypassing is accomplished by opening the internal joint between the two-in-hand sections in the faulted double pancake (internal pancake joint). In this way, the internal short circuit of the double pancake is disconnected completely from the other elements.

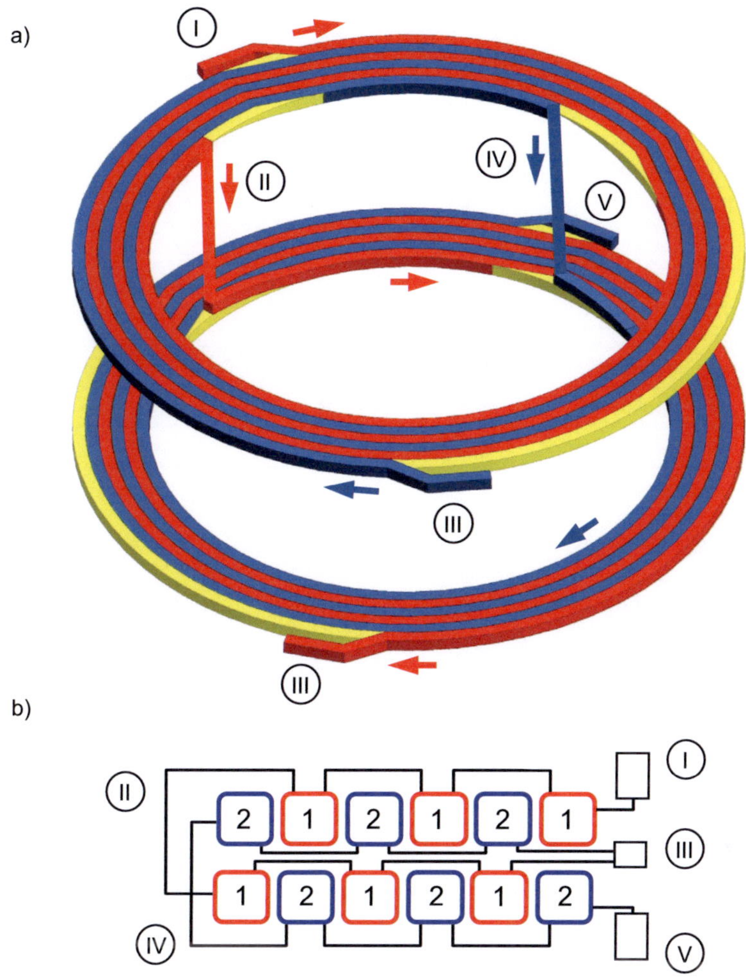

Fig. 2.7: Schematic concept of the two-in-hand double pancake winding configuration.
Part a: shows the winding configuration in one double pancake with current direction in conductors 1 and 2.
Part b: shows a cross-section of one double pancake with internal and external joints and turn transitions.
I.   External double pancake joint to next upper double pancake.
II. Transition of conductor 1 from upper pancake to lower pancake.
III. Internal joint between conductor 1 in lower pancake and conductor 2 in upper pancake.
IV. Transition of conductor 2 from upper pancake to lower pancake.
V.   External double pancake joint to next lower double pancake.

The total number of turns, the winding configuration in the PF coils and the length of the conductors are given in Tab. 2.4.

Tab. 2.4:   PF coil winding configuration [DDD06e].

PF coil	Conductor length	Conductor unit length	Number of pancakes	Number of turns	
				N_R x N_Z	Total
PF 1	6188 m	382 m	16	15.56 x 16	249
PF 2	5557 m	556 m	10	10.60 x 10	106
PF 3	13988 m	874 m	16	11.56 x 16	185
PF 4	12746 m	797 m	16	10.56 x 16	168
PF 5	11480 m	718 m	16	13.56 x 16	217
PF 6	11418 m	714 m	16	26.56 x 16	425

2.4   Power Supply Circuits

The CS and PF coils will be supplied with large pulsed electric power from a high-voltage grid. The circuits of the coil power supplies are designed to ensure continuous operation of all coils depending on the coil performance and requirements on the magnetic fields. The network model of the power supply circuits of CS and PF coils were used to calculate the voltage waveforms on the terminals of the PF 3 and PF 6 coils at the four specified calculation scenarios. The coil power supply circuits have three different designs, the single coil, the two coil series and the four coil parallel connection. Detailed information about power supplies of coils was taken from [PPS1].

The coils CS3U, CS3L, CS2U, CS2L, PF 1 and PF 6 have their own power supply circuits shown in Fig. 2.8. The 12-pulsed thyristor AC/DC main converter converts the alternating current from the high-voltage grid to the direct current needed for coil operation. Variations of coil voltage during normal operation required for plasma ignition and breakdown, which exceed the converter voltage, will be produced by the switching network unit (SNU). In case of a failure in the operation of ITER, e. g. quench in one of the superconducting coils, the fast discharge of all ITER coils will be started simultaneously by appropriate fast discharge units (FDUs) of the coil circuits. Each coil terminal is symmetrically grounded by the grounding resistor network, consisting of two terminal-to-neutral resistors R_{TN} and one neutral-to-ground resistor R_{NG}. This symmetrical grounding will limit the terminal to ground voltage of the coils to half, separated in positive and negative terminal voltage.

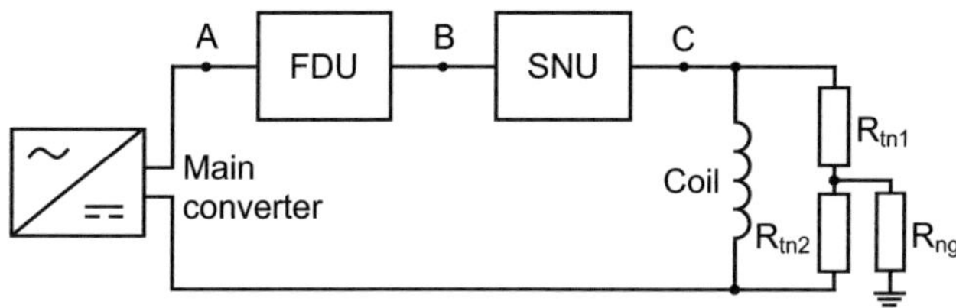

Fig. 2.8: Scheme of the coil power supply circuit for CS3L, CS2L, CS2U, CS3U coil modules, PF 1 and PF 6 coils. The circuit consists of a main converter, fast discharge unit (FDU), switching network unit (SNU), a coil and grounding resistor network (R_{tn1}, R_{tn2}, R_{ng}). The detailed design of the FDU between nodes A and B is shown in Fig. 2.11. The detailed design of the SNU between nodes B and C is shown in Fig. 2.12.

The CS1U and CS1L coil circuits are connected in series, which is shown in Fig. 2.9. They consist of the same elements as the single coil circuits, but with different values for the discharge resistors of the fast discharge units (FDU) of the CS1U and CS1L coils.

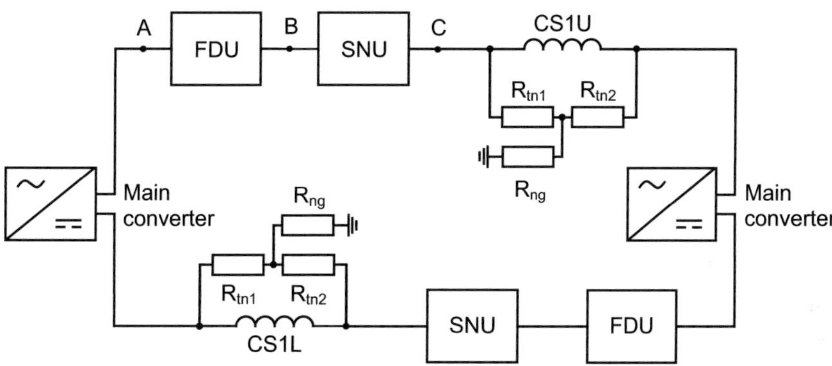

Fig. 2.9: Scheme of the coil power supply circuit for CS1L and CS1U coil modules. The circuit consists of two main converters, two fast discharge units (FDU), two switching network units (SNU), CS1L and CS1U coil modules and two grounding resistor networks (R_{tn1}, R_{tn2}, R_{ng}). The detailed design of the FDU between nodes A and B is shown in Fig. 2.11. The detailed design of the SNU between nodes B and C is shown in Fig. 2.12.

The PF 2, PF 3, PF 4 and PF 5 coils are connected in parallel in an electrical circuit, which is shown in Fig. 2.10. Due to requirements of fast control of the currents in these coils to stabilise the plasma vertical position, the vertical stabilisation converter is switched parallel to the coil circuits of PF 2 – PF 5. The fast discharge unit (FDU) and symmetrical resistor network for grounding have the same design as in other coil circuits, but with different values for the discharge resistor of the fast discharge unit. Instead of switching network units, booster converters are used in the circuits of the PF 2 – PF 5 coils. The voltage on the PF 2 – PF 5 coil terminals represents the cumulative voltage of the main converter (MC), booster converter (BC) and the vertical stabilisation converter (VSC) connected in series to each coil. Detailed drawings of the power supply circuits of each coil are shown in [PPSD1].

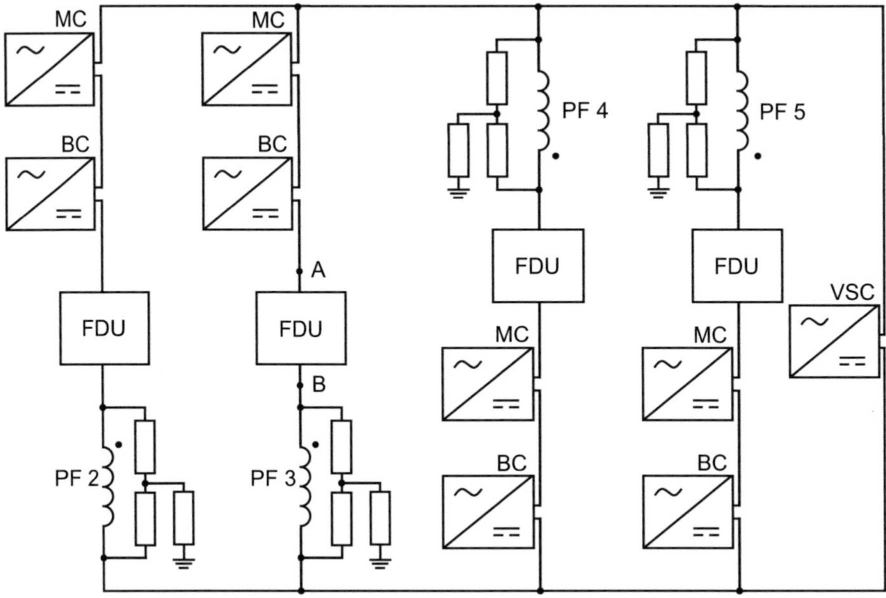

Fig. 2.10: Scheme of the coil power supply circuit for PF 2, PF 3, PF 4 and PF 5 coils. The circuit consists of four parallel identical coil circuit branches and a vertical stabilisation converter (VSC). Each branch consists of a main converter (MC), booster converter (BC), fast discharge unit (FDU), a PF coil and grounding resistor network (R_{tn1}, R_{tn2}, R_{ng}). The detailed design of the FDU between nodes A and B is shown in Fig. 2.11.

All CS and PF coils, except for the PF 2 coil, have a rated current of 45 kA. The rated current of the PF 2 coil is 41 kA. The main and vertical stabilisation converters have rated continuous output currents of 45 kA. The booster converters in the PF 2 – PF 5 coil circuits, by contrast, have a rated continuous output current 10 kA only. The main and booster converters are connected in series to the coils. Hence, the booster converters will be switched in the free wheeling mode, if the current of the respective branch will be higher than 10 kA. The no-load voltages, on-load voltages and rated continuous converter currents are summarised in Tab. 2.5.

Tab. 2.5:   Power supply converter data [PAF6Ba, PPS1a].

	Maximum no-load voltage	Maximum on-load voltage	Rated continuous output current
Main converter	2 kV	1.5 kV	45 kA
Booster converter	2 x 2.8 kV	2 x 2.1 kV	10 kA
Vertical stabilisation converter	2 x 4 kV	2 x 3 kV	45 kA

Fast Discharge Unit

The FDU will be used for fast discharge of the coils, if a failure occurs in the circuits, e. g. if a quench will be detected in one of the coils. The equivalent time constant for a fast discharge of CS coils is 11.5 s. It includes time for quench detection, 2 s action time for the switching procedure and current discharge time constant of 7.5 s [DDD06f]. For PF coils the equivalent time constant for fast discharge is 18 s, divided into time for quench detection, 2 s action time for the switching procedure and the current discharge time constant of 14 s [DDD06c]. Fig. 2.11 shows the schematic drawing of the fast discharge unit.

The FDU consists of a snubber circuit (SC), discharge resistor R_D, current commutation unit (CCU) and pyrobreaker (PB). The snubber circuit which will limit the high-frequency voltage oscillations caused by the FDU consists of a resistor with 0.15 Ω and a capacitor with 0.35 mF [PPSA1a]. The current commutation unit commutates the current from the free wheeling path to discharge resistor during fast discharge of the coils. The current commutation unit consists of a bypass switch (BPS), vacuum circuit breaker (VCB), saturable inductor L_S, counterpulse capacitor C_C and discharge thyristor switch Th. The pyrobreaker (PB) will be used to switch the current to the discharge resistor, if the current commutation unit will not work correctly.

At the beginning of fast discharge, the power supply of the coils will be disconnected from the coil circuit and a free wheeling path for coil current will be created by a switch. Then, the bypass switch will be opened and the coil current commutates to the parallel vacuum circuit breaker. After commutation to vacuum circuit breaker has been completed, it will be opened and an arc will occur inside the vacuum circuit breaker for some ms. To extinguish the arc in the vacuum circuit breaker, the counterpulse capacitor will be discharged by ignition of thyristor. The counterpulse capacitor with a capacitance of 1.8 mF will be charged to a voltage of 7 kV [PPSA1a] and stores enough energy for extinguishing of the arc in the vacuum circuit breaker. During discharge of counterpulse capacitor, the coil current through the vacuum circuit breaker will be extinguished and commutate to discharge resistor of the coil. The current commutation procedure is completed when the current through discharge resistor reaches the same value as the coil current.

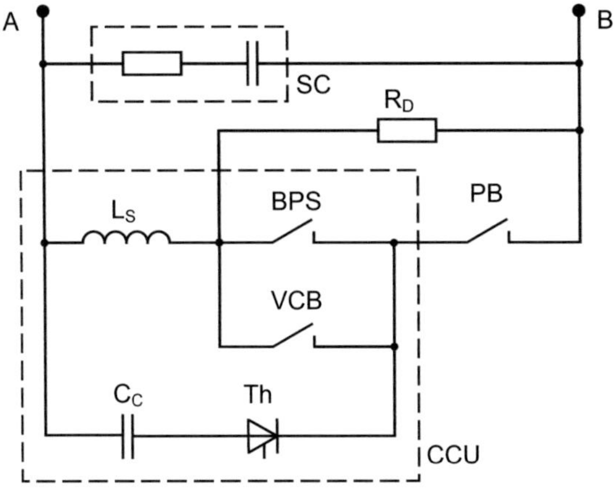

Fig. 2.11: Scheme of the fast discharge unit (FDU) consisting of a snubber circuit (SC) with resistor and capacitor, discharge resistor R_D, pyrobreaker (PB) and current commutation unit (CCU). The CCU consists of a saturable inductor L_S, bypass switch (BPS), vacuum circuit breaker (VCB), counterpulse capacitor C_C and thyristor Th. The nodes A and B show the connections of the FDU to power supply circuits of the coils shown in Fig. 2.8, Fig. 2.9 and Fig. 2.10.

The saturable inductor L_S of the fast discharge unit (FDU) is a non-linear current-depended inductor. The saturable inductor is used to limit the current variations at low current values and prevents the ignition of the current arc in the vacuum circuit breaker after crossing the 0 A level. The current dependence of the saturable inductor is shown in Tab. 2.6.

Tab. 2.6: Current dependence of the saturable inductor L_S for all fast discharge units (FDU) [PPSA1b].

Current	0 A	800 A	1200 A	1800 A	5000 A	20000 A
Inductance	156 µH	110 µH	77 µH	39 µH	10 µH	2 µH

During normal operation, the bypass switch and vacuum circuit breaker are closed and connected in parallel. The main part of the current flows through the bypass switch due to the 60 times higher resistance of the vacuum circuit breaker. The vacuum circuit breaker, by contrast, has a faster opening time and the electrodes of it can deal with an arc caused by opening of the vacuum circuit breaker during fast discharge. All relevant parameters of the bypass switch, vacuum circuit breaker and pyrobreaker are taken from [PPS1] and shown in Tab. 2.7.

Tab. 2.7: Parameters of the bypass switch (BPS), vacuum circuit breaker (VCB) and pyrobreaker (PB) used in all fast discharge units (FDU) [PPSA1b].

	BPS	VCB	PB
Resistance at main contacts	1 μΩ	60 μΩ	10 μΩ
Opening time	230 ms	30 ms	10 μs
Continuous current	68 kA	3.5 kA	68 kA
Pulsed current (duration)	250 kA (0.1 s)	68 kA (0.4 s)	100 kA (1 s)

Due to the different values of the stored energy in each coil, the discharge resistors have different values to reach the time constant specified for the coils. Relevant data of fast discharge resistors are summarised in Tab. 2.8. Further data of the fast discharge unit components can be taken from [PPS1, PPSA1].

Tab. 2.8: Data of fast discharge resistors [PAF6Bb].

Coils	Maximum FDU design energy	Initial value of fast discharge resistor at 20 °C	Maximum value for fast discharge resistor at 235 °C
CS3U	1.01 GJ	90 mΩ	212 mΩ
CS2U	1.15 GJ	103 mΩ	242 mΩ
CS1U	1.14 GJ	102 mΩ	240 mΩ
CS1L	1.13 GJ	101 mΩ	237 mΩ
CS2L	1.15 GJ	103 mΩ	242 mΩ
CS3L	1.15 GJ	91 mΩ	214 mΩ
PF 1	1.03 GJ	46 mΩ	108 mΩ
PF 2	0.42 GJ	19 mΩ	46 mΩ
PF 3	1.80 GJ	80 mΩ	188 mΩ
PF 4	1.51 GJ	67 mΩ	157 mΩ
PF 5	1.59 GJ	71 mΩ	167 mΩ
PF 6	2.07 GJ	92 mΩ	216 mΩ

Switching Network Unit

Terminal voltages which are necessary for plasma ignition and breakdown are too high to be delivered by main converters of the coil circuits. This additional voltage will be supplied by booster converters in each branch of the parallel circuit of the PF 2 – PF 5 coils, as shown in

Fig. 2.10. The additional voltage on terminals of all CS, PF 1 and PF 6 coils will be delivered by the switching network unit (SNU) shown in Fig. 2.8 and Fig. 2.9. The resistors of the switching network unit will be switched into the coil circuit to deliver the voltages needed for plasma ignition. The relatively high voltage at breakdown of plasma has to be provided at different currents between 40 % and 100 % of the nominal current. Therefore, several resistor branches are connected in parallel to reach the required voltage. The components of the switching network unit are shown in Fig. 2.12.

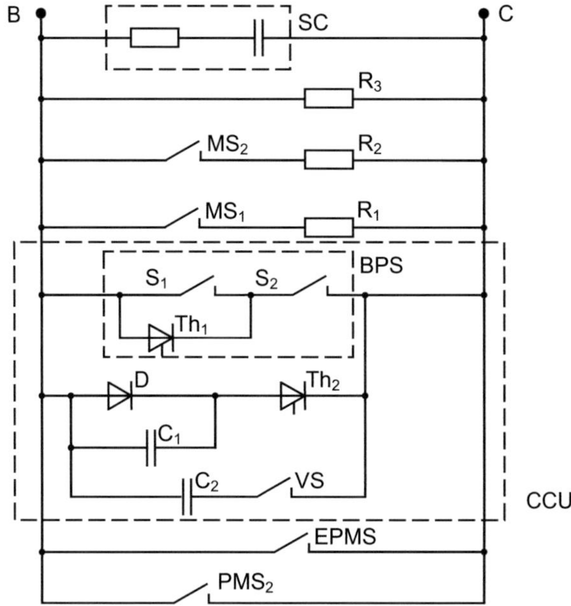

Fig. 2.12: Scheme of the switching network unit (SNU) consisting of a snubber circuit (SC), main switches 1 and 2 (MS$_1$ and MS$_2$), resistors 1, 2 and 3 (R$_1$, R$_2$ and R$_3$), explosively activated protective make switch (EPMS), protective make switch (PMS$_2$) and current commutation unit (CCU). The current commutation unit (CCU) consists of switches 1 and 2 (S$_1$ and S$_2$), thyristors 1 and 2 (Th$_1$ and Th$_2$), a diode D, capacitors 1 and 2 (C$_1$ and C$_2$), vacuum switch (VS) and a bypass switch (BPS). The nodes B and C show the connections of the switching network unit to the power supply circuits of the coils shown in Fig. 2.8 and Fig. 2.9.

The current commutation unit (CCU) is designed to avoid arcs during the commutation of current from bypass switch to switching network resistors. Additionally, the current commutation unit is used to bypass the switching network unit while charging the coils. The function of the current commutation unit is described in the section above. The current commutation unit of the switching network unit has two counterpulse capacitors C$_1$ and C$_2$ and a vacuum switch (VS). The capacitor C$_1$ with 20 mF will be charged to 1 kV and capacitor C$_2$ with 0.8 mF will be charged to

8 kV [PPSA1a]. The diode D and thyristor Th_2 are combined to a thyristor circuit breaker (TCB). The explosively activated protective make switch (EPMS) is connected in parallel with the current commutation unit to bypass the switching network unit in case of a failure in the current commutation unit. The snubber circuit (SC) is also placed in parallel to the current commutation unit and limits the high-frequency voltage oscillations caused by the switching network unit. The components of the snubber circuit are a capacitor with 0.1 mF and a resistor with 0.2 Ω [PPSA1a]. The protective make switches (PMS_1 and PMS_2) are built with a parallel bypass switch and vacuum circuit breaker. The protective make switch (PMS_1) will be used to connect the free wheeling circuit for coil current during fast discharge (not shown in Fig. 2.12). The protective make switch 2 will be used to bypass the switching network unit during fast discharge. The fast make switches (MS) will be used to switch additional resistances in parallel to achieve the required voltages during plasma ignition and breakdown. All relevant data of the switches used in the switching network unit are summarised in Tab. 2.9.

Tab. 2.9:   Parameters of the bypass switch (BPS), thyristor circuit breaker (TCB), fast make switch (MS) and explosively activated protective make switch (EPMS) used in the switching network unit [PPS1b].

	BPS	TCB	MS	EPMS
Rated current	60 kA	45 kA (for 10 ms)	60 kA	68 kA
Closing time	4.5 ms	n/a	0.1 ms	0.012 ms
Resistance at main contacts	5 $\mu\Omega$	n/a	1 $\mu\Omega$	3 $\mu\Omega$

The resistance network of the switching network unit consists of 9 modules switched in parallel to 3 module resistor groups (R_1, R_2 and R_3). The module resistor group R_3 is directly connected to the terminals of the switching network unit. The module resistor groups R_1 and R_2 will be switched parallel to main switches 1 and 2 to reduce the total resistance of the switching network unit. The rated values for the resistor modules of all CS and PF 1 coils are 10 Ω and for PF 6 coil 13.6 Ω [PPS1b].

3 Calculation Strategy

The internal voltage distribution of the PF 3 and PF 6 coils was calculated for different calculation scenarios. The excitations on the coil terminals at the calculation scenarios consist mainly of relatively slow current changes in the range of seconds and fast voltage excitations in the range of microseconds. The current and voltage behaviour on the coil terminals during fast discharge are shown using the simplified coil circuit in Fig. 3.1 as example. First, the superconducting coil L_S was driven in steady-state conditions with the coil current I_L and neglectable voltage between coil terminals. The discharge resistor R_D was bypassed by the free wheeling switch S_1. The fast discharge of the coil was started by the opening of the free wheeling switch and commutation of the current to the discharge resistor. The commutation causes a fast increase of the voltage between coil terminals in the range of microseconds. In the following, the current decreased with the current discharge time constant in the range of seconds depending on the inductance of the coil and the resistance of the discharge resistor.

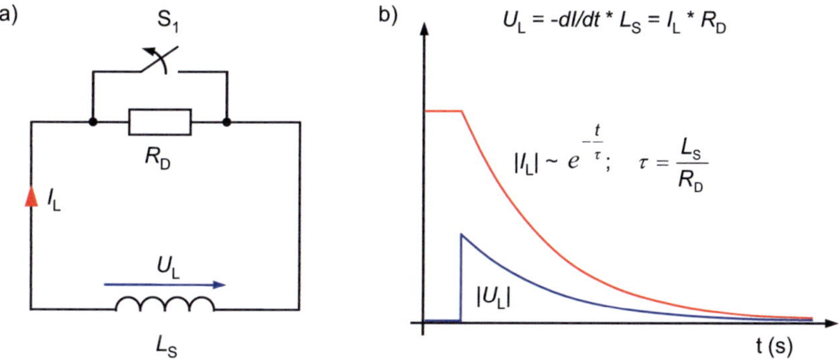

Fig. 3.1:   Part a: Simplified electrical circuit with superconducting coil L_S, discharge resistor R_D and free wheeling switch S_1.
Part b: The current and voltage behaviour on the coil terminals during fast discharge of the coil.

Similar behaviour of current in the range of seconds and voltage in the range of microseconds was calculated on the terminals of the PF coils. Additionally, the inductances of the PF coils are frequency-depended and decrease with increasing frequency. Thus, a special calculation strategy was applied for calculation of the transient electrical behaviour of ITER PF coils, shown in Fig. 3.2.

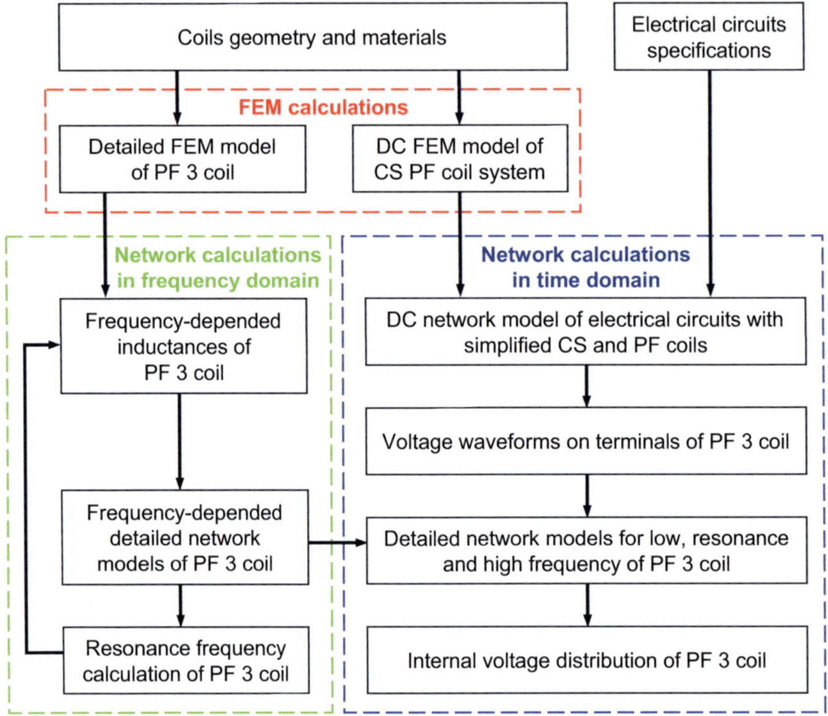

Fig. 3.2:   Strategy for calculation of resonance frequency and internal voltage distribution within the PF 3 coil. The strategy is shown for the PF 3 coil as an example and was used for the PF 6 coil, too.

The calculation strategy consists of three main parts. The first part includes the calculations of the frequency-depended inductances with a Finite Element Method (FEM). In this case, the FEM program Maxwell 2D [Max99] was used. The second part contains the calculations of resonance frequency in the frequency domain with the network program OrCAD PSpice [Orc07]. In the third part, the internal voltage distribution within PF 3 and PF 6 coils was calculated in the time domain with the network models which were also used for calculations of resonance frequencies. The FEM and network calculations were made with discrete frequency models. It has to be considered that the voltage excitations on the coil terminals consist of numerous frequencies in several frequency ranges with various voltage amplitudes. Additionally, the inductances of each turn of the PF coil are frequency-depended. Hence, the internal voltage distribution was analysed with models for low, resonance and high frequencies for each calculation scenario.

The two-dimensional rotation-symmetrical FEM model of the Central Solenoid (CS) and Poloidal Field (PF) coil system was built with the specified geometry and materials of CS and PF coils derived from [DDD06]. The self and mutual inductances of the CS and PF coils were calculated with the FEM model of the CS PF coil system at DC and used as lumped elements in a network model of this coil system. The detailed FEM models of the PF 3 and PF 6 coils were also built using the rotation-symmetry to calculate the frequency-depended self and mutual inductances for each turn of these coils at frequencies up to 300 kHz. The FEM models are described in detail in chapter 4. The frequency-independent capacitances between the turns were calculated with formulas for parallel plate and cylinder capacitors.

The network calculations in frequency domain included an iterative calculation loop and consisted of three steps. The FEM calculated frequency-dependent inductance and frequency-independent capacitance were taken as lumped elements and included in the detailed network models of the PF 3 and PF 6 coils. The network models are shown in detail in chapter 5. The calculations of the resonance frequency of the PF 3 and PF 6 coils were made with frequency-dependent detailed network models of both coils and provided the first benchmark for estimation of the transient behaviour of each coil. An iterative calculation loop was used for calculation of resonance frequency because of the frequency dependence of inductances of each turn which is described in detail in section 6.1. The detailed network models for different frequencies, which were built up for calculation of resonance frequency of the coils in the frequency domain were also used in some cases for calculation of the internal voltage distribution within the coils in the time domain.

The network calculations in time domain were performed to calculate the internal voltage distribution within PF 3 and PF 6 coils and consisted of four steps. At first, the DC network model of electrical circuits with simplified CS and PF coils was generated with values calculated with the DC FEM model of the CS PF coil system and the specifications of electrical circuits derived from [PPS1]. The voltage waveforms on the coil terminals of the PF 3 and PF 6 coils were calculated with the network model of the CS PF coil system for four calculation scenarios:

- Reference scenario describing the current behaviour of CS and PF coils during normal operation.
- Fast discharge of coils, e. g. in case of a quench of one of the superconducting coils.
- Failure case 1 defined as an earth fault on negative coil terminal during fast discharge. The coils will be driven at medium DC voltages and rated currents.
- Failure case 2 defined as an earth fault on negative coil terminal during fast discharge. The coils will be driven at rated DC voltages and medium currents.

The four calculation scenarios with relevant current and voltage waveforms are described in detail in section 6.2.

Three detailed network models for each the PF 3 and the PF 6 coil were derived from several detailed network models built for different frequencies in calculations of resonance frequency. The first network model was generated for low-frequency excitations similar to the excitations calculated during the reference scenario. The second network model was designed for excitations which were close to resonance frequency of each coil, comparable with excitations calculated in both failure cases. The third network model was for high-frequency excitations close to the excitations calculated for fast discharge and both failure cases. All three network models of each coil were excited by voltage waveforms calculated for the four calculation scenarios. The results of the calculations for each scenario were compared with each other to analyse the influence of the frequency dependence of inductances on the behaviour of internal coil voltages. The voltage waveforms and maximum voltages were calculated on the ground, layer and turn insulation of PF 3 and PF 6 coils and are described in detail in section 6.3.

4 Finite Element Method Calculations

The Finite Element Method (FEM) model of the CS PF coil system was set up to calculate self and mutual inductances of CS and PF coils at DC due to the change of coil currents in the range of seconds. The definition of FEM model and calculation results are presented in section 4.1. The calculated values will be used as lumped elements in the network model of the coil power supply circuits.

The detailed FEM models of the PF 3 and PF 6 coils were set up to calculate the frequency-depending self and mutual inductances of each turn in these coils for frequencies up to 300 kHz. The detailed FEM models of the PF 3 and PF 6 coils and calculation results are described in sections 4.2 and 4.3, respectively. The calculated inductances will be used as lumped elements in detailed network models of the PF 3 and PF 6 coils. Additionally, the values for superconducting and water-cooled bus bar were calculated with their FEM models and are described in section 4.4.

4.1 Simplified FEM Model of the ITER Coil System

The rotational symmetry of the coil system was used to set up two-dimensional models of the coils using the FEM program Maxwell 2D [Max99]. At first, a simplified model of the CS PF coil system was set up, which is shown in Fig. 4.1. The coils are shown as rectangles with the cross-sections of the CS and PF coils. The positions of the coil centres and dimensions in Z- and R-directions are summarised in Tab. 2.1.

The self and mutual inductances of simplified models of the CS and PF coils were calculated with FEM at DC. Equal current density was defined in the rectangles which were equivalent to the cross-sections of the CS and PF coils. Vacuum was chosen for the space between the coils. The coil cross-sections were defined to be a stranded conductor, which means that the conductor is transposed thus skin and proximity effects are neglected in these cross-sections.

The boundary of the model was set to ±30 m in Z-direction and to 40 m in R-direction. The boundaries were defined to be magnetically neutral with magnetic zero potential which was infinitely far away from the model. The currents in the CS and PF coils were set to 45 kA and used by the FEM program for calculation of magnetic fields only. To calculate the self and mutual inductances, the currents through the coils were automatically set to 1 A by the FEM program. Triangular elements were used for meshing of the model. Appropriate meshing leads to accurate calculation results and shortens the calculation time. Not only the number of the meshing elements is important, but also their location and size in the model. A difference of 1 % in energy between the meshing triangles was chosen as a criterion for appropriate meshing. This means that the difference between the energies of the magnetic field stored in different meshing triangles was smaller than 1 %. Thus, the mesh in regions with a high magnetic field strength had a smaller size than the mesh in regions with a low magnetic field strength. Detailed information about the setting up of the FEM model of the CS PF coil system in Maxwell 2D is summarised in Annex A.1.

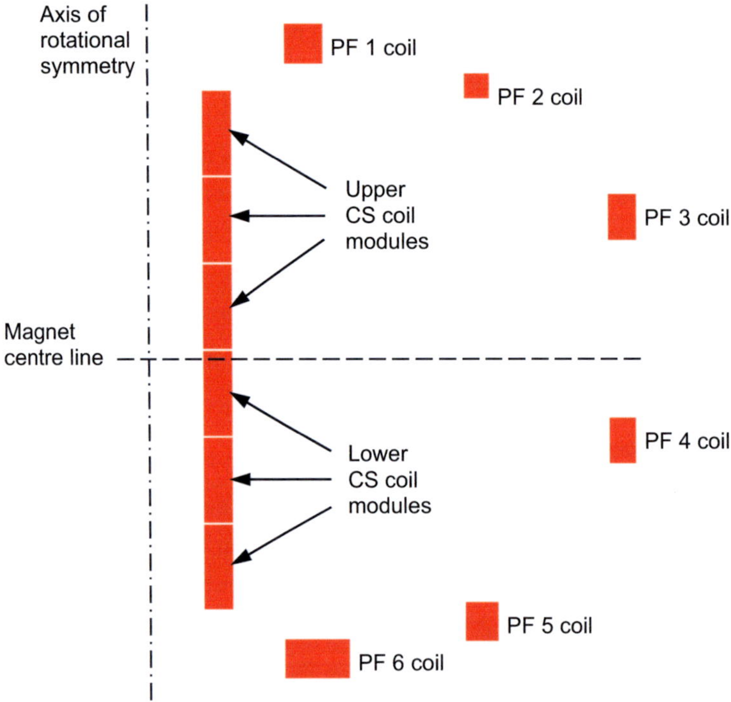

Fig. 4.1:   Simplified two-dimensional FEM model of the ITER CS PF coil system in Maxwell 2D.

The inductance of each cross-section L_{CS} of the modelled coils was calculated at DC. To obtain the self inductance of the coils L_C, the calculated inductance of coil cross-section L_{CS} has to be multiplied by the squared number of turns of the coil w, according to formula 4.1.

$$L_C = w^2 \cdot L_{CS}$$

[4.1]

The calculated self inductances of the CS and PF coils are summarised in Tab. 4.1. Due to the small coil diameter of the CS coil modules compared to the coil diameters of the PF coils, the cross-sections of CS coil modules have a relatively low inductance. However, CS coil modules have a larger number of turns than PF coils, thus the self inductance of CS coil modules is higher than the self inductance of PF 1 and PF 2 coils, despite the lower inductance of the cross-section of CS coil modules.

Tab. 4.1: FEM-calculated DC values for self inductances with simplified CS and PF coil models.

	CS module	PF 1	PF 2	PF 3	PF 4	PF 5	PF 6
FEM-calculated inductance of coil cross-section (μH)	2.59	11.38	36.35	53.42	53.79	33.1	11.16
Number of turns	549	249	106	185	168	217	425
DC self inductance of the coil (H)	0.778	0.705	0.408	1.828	1.536	1.559	2.016

The calculation of mutual inductances between CS and PF coils $M_{C1\text{-}C2}$ was started with the calculation of mutual inductances between the cross sections of the coils $M_{CS1\text{-}CS2}$ in FEM program. The mutual inductance of the coils $M_{C1\text{-}C2}$ was calculated by multiplying of mutual inductance between the coil cross-sections $M_{CS1\text{-}CS2}$ by the number of turns of the first coil w_{C1} and one time with the number of turns of the second coil w_{C2}, according to formula 4.2. To calculate the mutual inductance between CS1U and PF 3 coils, for example, the mutual inductance calculated by Maxwell 2D was multiplied by the number of turns of the CS1U coil module and with the number of turns of the PF 3 coils.

$$M_{C1-C2} = w_{C1} \cdot w_{C2} \cdot M_{CS1-CS2} \qquad [4.2]$$

Mutual inductance depends on the distance between the coils, number of turns and coil geometry. The calculated mutual inductances between CS and PF coils at DC are given in Tab. 4.2 and Tab. 4.3.

Tab. 4.2:   Mutual DC inductance between the CS coil modules and PF coils.

	CS3U	CS2U	CS1U	CS1L	CS2L	CS3L
CS2U	0.241 H					
CS1U	0.052 H	0.241 H				
CS1L	0.018 H	0.052 H	0.241 H			
CS2L	0.008 H	0.018 H	0.052 H	0.241 H		
CS3L	0.004 H	0.008 H	0.018 H	0.052 H	0.241 H	
PF 1	0.136 H	0.061 H	0.028 H	0.014 H	0.008 H	0.005 H
PF 2	0.041 H	0.033 H	0.024 H	0.017 H	0.011 H	0.008 H
PF 3	0.048 H	0.050 H	0.048 H	0.042 H	0.034 H	0.027 H
PF 4	0.028 H	0.035 H	0.041 H	0.045 H	0.046 H	0.042 H
PF 5	0.015 H	0.022 H	0.033 H	0.048 H	0.066 H	0.081 H
PF 6	0.010 H	0.016 H	0.028 H	0.052 H	0.108 H	0.227 H

Tab. 4.3:   Mutual DC inductance between the PF coils.

	PF 1	PF 2	PF 3	PF 4	PF 5
PF 2	0.104 H				
PF 3	0.100 H	0.220 H			
PF 4	0.049 H	0.094 H	0.448 H		
PF 5	0.024 H	0.043 H	0.182 H	0.358 H	
PF 6	0.014 H	0.024 H	0.093 H	0.166 H	0.437 H

4.2   Detailed FEM Model of the Poloidal Field 3 Coil

For the calculation of the internal voltage distribution within the PF 3 and PF 6 coils, the values for elements of the network models were determined using the detailed FEM models of these coils. The setting up of the detailed FEM model of the PF 3 coil at 4.5 K is described in this chapter. The dimensions and materials of the detailed FEM model of the coil are shown in the first section. The results of the FEM calculations for the PF 3 coil are discussed in the second section.

4.2.1   Description of the FEM Model

Values for frequency dependent-inductances were calculated with the detailed FEM model for each turn of the PF 3 coil. The detailed FEM model of the PF 3 coil is shown in Fig. 4.2. The coil consists of 16 layers with 12 turns in each layer. A single turn is represented by a superconducting cable surrounded by a stainless steel jacket and a cooling tube inside. Electrical insulation is placed between the turns and layers.

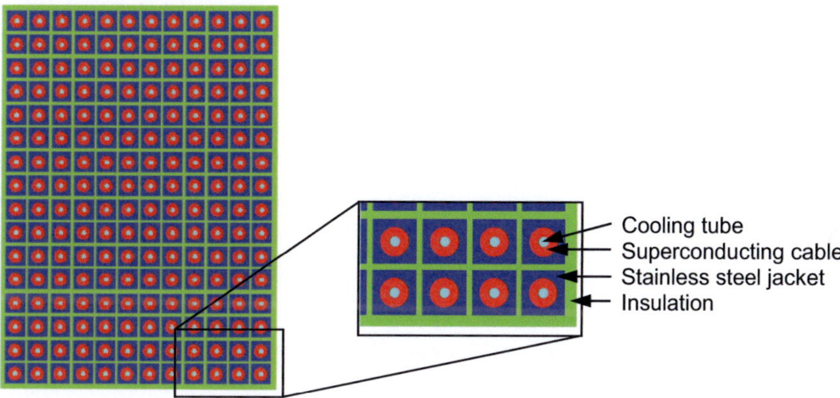

Fig. 4.2:   Detailed FEM model of a cross-section of the PF 3 coil in Maxwell 2D. The axis of rotational symmetry (not shown in the figure) is located to the left of the cross-section. The zoomed area shows the outer lower corner of the PF 3 coil model.

The outer diameter of the cooling tube is 12 mm. The thickness of the cooling tube wall, its metallic properties and properties of the helium inside the cooling tube were neglected for the calculations of outer inductances and capacitances, because the cooling tube is placed inside the current-leading superconducting cable and, thus, has no influence on the outer magnetic and electric fields of the conductor. The definition of the cooling tube as a vacuum creates a current-free space inside the superconducting cable, which is important to the calculation of the inner inductance of the conductor.

The superconducting cable of the PF 3 coil consists of 864 transposed superconducting strands. Consequently, the superconducting cable in FEM was defined to be a stranded conductor, which means that the conductor is transposed, thus, skin and proximity effects will not occur in the superconducting cable and a uniform current density is defined over the complete cable cross-section. To calculate the inductance, copper with a conductivity of 6.4e9 S/m at 4.5 K [DRGAa] was chosen as the material for the superconducting cable in the FEM models. The outer diameter of the superconducting cable of 34.4 mm was calculated from cable dimensions at room temperature [DDDD1] multiplied by the shrinkage factor of 0.29 % [DDD06g] at cooling down to 4.5 K. Due to the stranded structure of the superconducting cable, the stainless steel jacket has a strong influence on the cable dimensions at 4.5 K. Hence, the diameter of the superconducting cable also shrinks by 0.29 %. The stainless steel protection wrap of 0.08 mm with 50 % overlap [DDD06c] is used to protect the cable during the pulling processes through the stainless steel jacket. The thickness of the protection wrap was added to the stainless steel jacket in the detailed FEM models of the PF 3 coil.

The stainless steel jacket is made for the mechanical stability of the superconducting cable during the operation of the coils. The outer dimensions of the jacket were 52.1 mm, calculated with a shrinkage factor of 0.29 %. The dimensions at room temperature are given in Fig. 2.4. The conductivity of stainless steel was set to 1.88e6 S/m at 4.5 K, which is given in [DRGAa].

The different kinds of insulation, e. g. turn, layer and ground insulation, were summarised to one insulation type because of their similar permittivity of about $\varepsilon_r = 4$ [DRG1a]. Each conductor has a 3.0 mm turn insulation. Furthermore, an adjustable turn insulation of 0.5 mm is installed between two turns of the same pancake. Hence, the total turn-to-turn insulation is 6.5 mm. The insulation in Z-direction depends on the location of the conductors in the coil. The adjustable layer insulation in one double pancake is 1 mm. Hence, the total layer insulation between two layers in one double pancake is 7.0 mm. An additional insulation layer called bounding layer is placed between the double pancakes and has a thickness of 2 mm. The total layer insulation between the conductors of two different double pancakes is 9.0 mm [DDDD2].

The ground insulation of 8 mm and outer protection layer of 2 mm, which cover all turns of the PF 3 coil, were summarised to an outer insulation with a thickness of 10 mm. The total dimensions of the PF 3 coil with ground insulation are 728 mm in R-direction and 986 mm in Z-direction. Compared to the coil dimensions without ground insulation of 708 mm and 966 mm given in Tab. 2.1 for 4.5 K, the 10 mm of ground insulation on both sides of the coil were considered in the detailed FEM model of the PF 3 coil.

The detailed FEM model of the PF 3 coil shown in Fig. 4.2 was used for FEM calculations at DC. With increasing frequencies of the FEM models, two additional FEM models of the PF 3 coil were used, because the models at higher frequencies need more working memory than Maxwell 2D's maximum of 2 GB. At frequencies higher than DC and lower than 5 kHz, a FEM model of half of the PF 3 coil with even line symmetry was used, which is shown in Fig. 4.3. Using the even line symmetry in Maxwell 2D allows to draw the upper half of the whole FEM model only. With the definition of the even line symmetry, the lower part of the model will be considered in the calculations of the magnetic fields and also in the impedance matrix. The function and performance of even line symmetry were tested successfully with the DC models of the complete and half FEM models of the PF 3 coil before the beginning of calculations at frequencies up to 5 kHz. The only difference between the FEM model shown in Fig. 4.2 and Fig. 4.3 is that the model in Fig. 4.3 corresponds to the half of the PF 3 coil. All other dimensions of the coil remain the same.

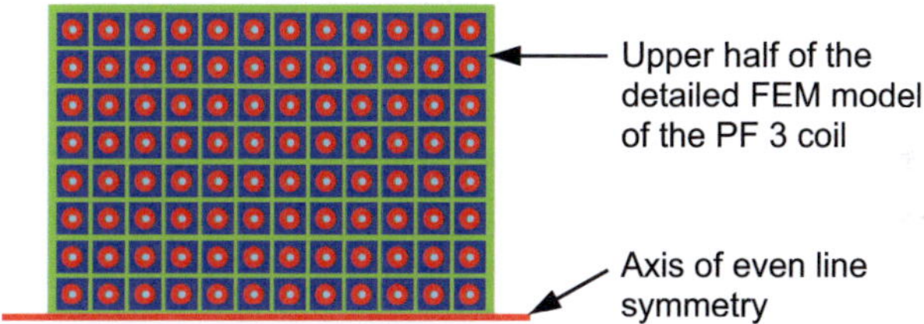

Fig. 4.3:   FEM model of a cross-section of half of the PF 3 coil with even line symmetry axis at the bottom. The axis of rotational symmetry is to the left (not shown in the figure).

With increasing frequency, the conductor inductance decreases due to the higher eddy currents induced in the stainless steel jacket of each conductor. The induction of eddy currents in the stainless steel jacket means that both the self inductance of one conductor and mutual inductance between two conductors decrease. For frequencies higher than 5 kHz, the magnetic coupling between the conductors is lower than 1 % of the self inductance of this conductor. Thus, mutual inductance between the conductors was neglected. For frequencies higher than 5 kHz, the FEM model of only one pancake of the PF 3 coil was used, which is shown in Fig. 4.4.

Fig. 4.4:   FEM model of a cross-section of one pancake of the PF 3 coil.

The dimensions of the conductors and insulation were the same as in the models in Fig. 4.2 and Fig. 4.3. The values for the inductance of each conductor in one pancake were calculated and used for other pancakes of the PF 3 coil.

4.2.2   Calculation of Frequency-Depended Inductances

This section will present a part of the FEM results only, to explain different mechanisms of the frequency dependence of the inductances of the PF 3 coil. The detailed results of the FEM calculations of the PF 3 coil are summarised in Annex A.1.

The inductance of a conductor is the factor reflecting the dependence between the current of the conductor and the magnetic energy stored in the space around and inside the conductor. The self inductance of each conductor is the sum of the inner and the outer self inductances. The inner self inductance defines the dependence between the current and the magnetic energy stored in the space inside the conductor. The outer self inductance is the relation between the current and the magnetic energy stored in the space around the conductor. The inner self inductance of a conductor remains the same over the whole frequency range, if the conductor is defined as transposed and skin and proximity effects do not occur in this conductor. The superconducting cables used in PF coils consist of many transposed thin filaments, whose dependence on skin and proximity effects can be neglected. Thus, the inner self inductance of the superconducting cable remains the same in the relevant frequency range up to 300 kHz.

The outer self inductance of the conductor depends on the design of the conductor and is strongly frequency-dependent in case of PF coils. The eddy currents induced in the stainless steel jacket of the conductor decrease the magnetic field around the conductors. Consequently, the magnetic energy used for the calculation of the outer self inductance decrease as well. With increasing frequency, the eddy currents in the stainless steel jacket increase as well and, thus, the outer self inductance of the conductor decreases.

The frequency dependence of the self inductance of the upper left conductor (i. e. turn) is shown in Tab. 4.4. The upper left conductor has the number P01_C01. The first part stands for the pancake number from top to bottom of the coil, the second denotes the conductor number from the inner to the outer side of the coil. So the next conductor in the same pancake has the number P01_C02, the next P01_C03 and so on. The first left conductor of the next pancake has the number P02_C01. The calculated self inductances of the conductor were taken from the calculations made for the frequencies relevant to the calculations of resonance frequency of the PF 3 coil shown in section 6.1.

The strongest drop of self inductance took place between DC and 0.50 kHz, with the factor being 11.9, as shown in Tab. 4.4. At 0.50 kHz, the eddy currents induced in the stainless steel jacket which weaken the outer magnetic field of the conductor, are very strong compared to the situation without any eddy currents at DC. The outer self inductance decreased with increasing frequency and at 300 kHz, only the inner self inductance of the conductor is relevant to the calculation of self inductance.

Tab. 4.4:   Frequency dependence of self inductance of conductor (i. e. turn) P01_C01 of the PF 3 coil shown for frequencies used for calculation of resonance frequency of the coil.

Frequency	0 kHz	0.50 kHz	10.0 kHz	25.5 kHz	200 kHz	300 kHz
Self inductance of the turn P01_C01	9.9e-05 H	8.3e-06 H	4.5e-06 H	4.0e-06 H	3.3e-06 H	3.3e-06 H
Relative factor to the self inductance of the turn P01_C01 at DC	1	0.084	0.045	0.040	0.034	0.033

The cable-in-conduit design of the PF coil conductor also has a strong influence on the frequency dependence of mutual inductance between the turns shown in Tab. 4.5. The eddy currents induced in the stainless steel jacket of the conductors of the PF coils weaken the mutual inductance between the turns. With increasing frequency, mutual inductance between two turns decreases. It continues to decrease also at very high frequencies contrary to the self inductance of the turn, which reaches a convergence value of inner inductance at a frequency of about 300 kHz.

Tab. 4.5:   Mutual inductance between the turn P01_C01 and P01_C02 of the PF 3 coil shown for frequencies used for calculation of the resonance frequency of the coil.

Frequency	0 kHz	0.50 kHz	10.0 kHz	25.5 kHz	200 kHz	300 kHz
Mutual inductance between turn P01_C01 and P01_C02	7.8e-05 H	1.2e-06 H	1.0e-09 H	4.8e-11 H	1.2e-18 H	2.2e-19 H
Coupling factor between turn P01_C01 and P01_C02	0.79	0.14	3e-04	1e-05	4e-18	7e-19

With increasing distance between the turns, their mutual inductance decreases. At DC, this effect is less pronounced, for example, mutual inductance between conductors P01_C01 and P01_C02 is 7.8e-05 H and between P01_C01 and P01_C12 4.4e-05 H, the factor between these values being 1.76 only. Selected mutual inductances between conductors of the upper pancake at DC are shown in Tab. 4.6. For a frequency of 300 kHz, mutual inductance between conductor P01_C01 and P01_C02 is 2.2e-19 H and between P01_C01 and P01_C12 7.6E-22 H only, the factor between these values is 294. Selected mutual inductances between turns of pancake 01 at 300 kHz are shown in Tab. 4.7.

Tab. 4.6:   Selected DC mutual inductances and coupling factors between the turns in pancake 01 of the PF 3 coil.

Turn number	P01_C02	P01_C04	P01_C06	P01_C08	P01_C10	P01_C12
Mutual inductance with turn P01_C01	7.8e-05 H	6.2e-05 H	5.5e-05 H	5.0e-05 H	4.7e-05 H	4.4e-05 H
Coupling factor with turn P01_C01	0.79	0.63	0.55	0.50	0.46	0.43

Tab. 4.7:   Selected mutual inductances and coupling factors between the turns in pancake 01 of the PF 3 coil at 300 kHz.

Turn number	P01_C02	P01_C04	P01_C06	P01_C08	P01_C10	P01_C12
Mutual inductance with turn P01_C01	2.2e-19 H	4.2e-21 H	9.8e-22 H	1.4e-21 H	9.0e-22 H	7.6e-22 H
Coupling factor with turn P01_C01	7e-14	1e-15	3e-16	4e-16	2e-16	2e-16

Cumulative inductance of each turn was calculated by adding all mutual inductances of a turn with all other turns of the PF 3 coil and the self inductance of this turn. Cumulative inductance of the turn P01_C01 is 8.9e-3 H at DC. This value is very high compared to the self inductance of the same turn of 9.88e-05 H. This high cumulative inductance is the result of the high values of the mutual inductances between the turns at DC, which are shown in Tab. 4.6. With increasing frequency, mutual inductance decreases faster than self inductance, which is shown in Tab. 4.4 and Tab. 4.5. At frequencies higher than 5 kHz, mutual inductances between the turns were neglected, because their values were lower than 1 % of the self inductance and only self inductance of the turns was taken into account. The frequency dependence of the cumulative inductance of the turn P01_C01 is shown in Tab. 4.8.

Tab. 4.8:   Frequency dependence of cumulative inductance of turn P01_C01 of the PF 3 coil.

Frequency	0 kHz	0.50 kHz	10.0 kHz	25.5 kHz	200 kHz	300 kHz
Cumulative inductance of turn P01_C01	8.9e-03 H	1.0e-05 H	4.5e-06 H	4.0e-06 H	3.32e-06 H	3.26e-06 H

The total inductance of the PF 3 coil calculated with the detailed FEM model, was compared with the inductance of the PF 3 coil calculated with the simplified model of the CS PF coil system. Total inductances were compared at DC only, because the inductance calculation with the simplified model of the coil system had been made for DC, exclusively. The total inductance of the detailed FEM model of the PF 3 coil was calculated by the sum all cumulative inductances of each turn. Further the self inductance of the PF 3 coil was calculated with the formula given in [Gro62] for DC and compared with ITER coil specifications derived from [PAF6Bd]. The calculated total inductances of the PF 3 coil with four calculation methods are summarised in Tab. 4.9.

Tab. 4.9: Comparison of the cumulative total inductance of the PF 3 coil calculated at DC with different methods.

Calculation method	PF 3	Relative difference to detailed FEM model
Detailed FEM model	1.94 H	-
Simplified FEM model	1.83 H	5.6 %
Formula from [Gro62]	1.80 H	7.2 %
ITER specification [PAF6Bd]	1.78 H	8.2 %

These relatively small differences show that calculation of total inductance with the formula given in [Gro62] and with the simplified FEM model of a coil can be used to estimate self inductance of big superconducting coils. More detailed calculations of inductances and field analysis of the coil were made with the detailed FEM models.

4.3   Detailed FEM Model of the Poloidal Field 6 Coil

The frequency-dependent inductances of the PF 6 coil were calculated with the same strategy as used for the PF 3 coil. Nevertheless, all relevant information for the setting up of the detailed model of the PF 6 coil is described in detail in this chapter. The dimensions and materials of the detailed FEM model of the PF 6 coil are presented in the first section. The results of the FEM calculation for the PF 6 coil are discussed in the second section.

4.3.1   Description of the FEM Model

Frequency-dependent inductances for each turn were calculated with the detailed FEM model of the PF 6 coil. The FEM model of the PF 6 coil consists of the cooling tube, superconducting cable, stainless steel jacket and insulation, which are shown in Fig. 4.5.

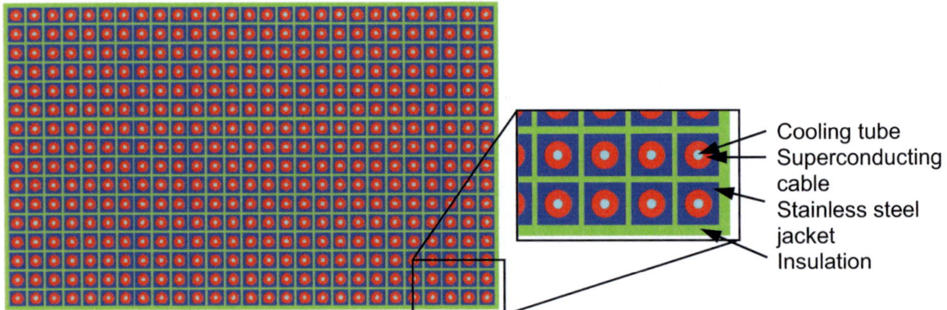

Fig. 4.5:   Detailed FEM model of a cross-section of the PF 6 coil in Maxwell 2D. The axis of rotational symmetry (not shown in the figure) located to the left of the cross-section. The zoomed area shows the outer lower corner of the PF 6 coil model.

The outer diameter of the cooling tube is 12 mm. The thickness of the cooling tube wall, its metallic properties and the properties of the helium inside the cooling tube were neglected in the calculations of outer inductances and capacitances, because the cooling tube is placed inside the current-leading superconducting cable and, thus, has no influence on the outer magnetic and electric fields of the conductor. The definition of cooling tube as a vacuum creates a current-free space inside the superconducting cable with equal current distribution which is important to the calculation of the inner inductance of the conductor.

The superconducting cable of the PF 6 coil consists of 1440 superconducting strands. Hence, the superconducting cable in FEM was defined to be stranded conductor, which means that the conductor is transposed. As a result, the skin and proximity effects in the cable are neglected and uniform current density is defined over the whole cable cross-section. Copper with the conductivity of 6.4e9 S/m at 4.5 K [DRGAa] was chosen as the material for the superconducting cable in the FEM models. The outer diameter of the superconducting cable of 38.1 mm was calculated from dimensions of the cable at room temperature [DDDD1] multiplied by the shrinkage factor of 0.29 % [DDD06g] when cooling down to 4.5 K. Due to the stranded structure of the superconducting cable, the stainless steel jacket has a strong influence on the cable dimensions at a temperature of 4.5 K. Consequently, the diameter of the superconducting cable also shrinks 0.29 %. The stainless steel protection wrap of 0.08 mm with 50 % overlap [DDD06c] is used to protect the cable when it is pulled through the stainless steel jacket. The thickness of the protection wrap was added to the stainless steel jacket in the FEM model of the PF 6 coil.

The stainless steel jacket is used to give the superconducting cable mechanical stability during the manufacturing process and operation of the coils. The outer dimensions of the jacket are 53.6 mm at 4.5 K, calculated with a shrinkage factor of 0.29 % compared to dimensions at room temperature, which are given in Fig. 2.4. The conductivity of stainless steel at 4.5 K was set to 1.88e6 S/m according to [DRGAa].

The different kinds of insulation, e. g. turn, layer and ground insulation, were summarised to one insulation type because of the similar permittivity of about $\varepsilon_r = 4$ [DRG1a]. Each conductor has a 3.0 mm thick turn insulation. Furthermore, an adjustable turn insulation of 0.5 mm is installed between two turns in the same pancake of the coil. Thus, the total turn-to-turn insulation is 6.5 mm. The insulation in Z-direction depends on the location of the conductors in the coil. The adjustable layer insulation in one double pancake is 0.5 mm. Thus, the total layer-to-layer insulation in one double pancake is 6.5 mm. An additional insulation layer called bounding layer is placed between the double pancakes and has a thickness of 1 mm. The total layer-to-layer insulation between the layers of two different double pancakes is 7.5 mm [DDDD2].

The ground insulation of 8 mm and outer protection layer of 2 mm, which cover all turns of the PF 6 coil, were summarised to an outer insulation with a thickness of 10 mm. The total dimensions of the PF 6 coil with ground insulation are 996 mm in Z-direction and 1653 mm in R-direction. Compared with the coil dimensions of 976 mm and 1633 mm specified for 4.5 K without ground insulation in Tab. 2.1, there are 10 mm of ground insulation on both sides of the coil, which were considered in the detailed FEM model of the PF 6 coil.

The detailed FEM model of the PF 6 coil shown in Fig. 4.5 was used for FEM calculations at DC. With increasing frequencies of the FEM models, two additional FEM models of the PF 6 coil were used, because the models at higher frequencies need more working memory than Maxwell 2D's maximum of 2 GB. At frequencies higher than DC and lower than 5 kHz, a FEM model of the PF 6 coil with even line symmetry was used, which is shown in Fig. 4.6. Using the even line symmetry in Maxwell 2D allows to draw the upper half of the whole FEM model only. With the definition of the even line symmetry, the lower part of the model is considered in calculations of magnetic fields and the impedance matrix. The function and performance of even line symmetry had been tested successfully with the DC models of the complete and half FEM model of the PF 3 coil before the beginning of calculations at frequencies up to 5 kHz. The only difference between the FEM models shown in Fig. 4.5 and Fig. 4.6 is that the model in Fig. 4.6 presents half of the cross-section of the PF 6 coil. All other dimensions of the coil remain the same.

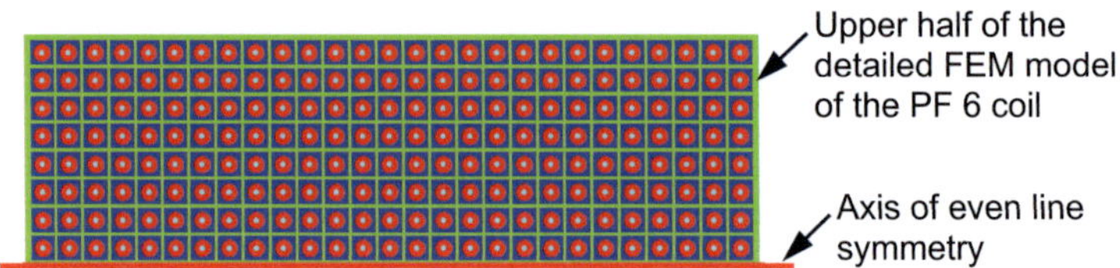

Fig. 4.6:   FEM model of a cross-section of half of the PF 6 coil with even line symmetry axis at the bottom. The axis of rotational symmetry is to the left (not shown in the figure).

With increasing frequency of the calculations, the inductance of the conductor decreases due to the higher eddy currents induced in the stainless steel jacket of each conductor. The induction of eddy current in the stainless steel jacket means that both the self inductance of a conductor and the mutual inductance between two conductors decreases. For frequencies higher than 5 kHz, magnetic coupling between the conductors is lower than 1 % of the self inductance of the conductor. Mutual inductance between the conductors was therefore neglected. At frequencies higher than 5 kHz, the FEM model of only one pancake of the PF 6 coil was used, which is shown in Fig. 4.7. The dimensions of the conductors and insulation were the same as in the models in Fig. 4.5 and Fig. 4.6. The values for the inductance of each turn in one pancake were calculated and used for other pancakes.

Fig. 4.7:   FEM model of one pancake of the PF 6 coil.

4.3.2   Calculation of Frequency-Depended Inductances

This section presents a part of the results only to explain the different mechanisms of frequency dependence of the inductances of the PF 6 coil. The detailed results of the FEM calculations of the PF 6 coil are summarised in Annex A.1.

The inner inductance of the conductor remains the same over the whole frequency range, if no skin and proximity effects occur in the conductor. Due to the transposed multi-filament design, the superconducting cable has small dependence on skin and proximity effects. Hence, inner inductance is not reduced in the relevant frequency range up to 300 kHz. On the other hand, the eddy currents induced in the stainless steel jacket of the conductor decrease the magnetic field around the conductors and also the magnetic energy used for calculating outer inductance. With increasing frequency, the eddy currents in the stainless steel jacket increase as well and outer inductance of the conductor decreases. The frequency dependence of the self inductance of the upper left conductor (i. e. turn) having the number P01_C01 is shown in Tab. 4.10. The first part of the designation stands for the pancake number from the top to the bottom of the coil, the second for the conductor number from the inner to the outer side of the coil. The values for the self inductance of the conductor were taken from the calculations made for the frequencies relevant to the calculation of resonance frequency of the PF 6 coil shown in section 6.1.

Tab. 4.10: Frequency dependence of self inductance of the conductor (i. e. turn) P01_C01 of the PF 6 coil shown for frequencies used for calculation of the resonance frequency of the coil.

Frequency	0 kHz	0.5 kHz	13 kHz	32 kHz	200 kHz	300 kHz
Self inductance of the turn P01_C01	2.4e-05 H	2.3e-06 H	1.2e-06 H	1.1e-06 H	1.0e-6 H	9.9e-7 H
Relative factor to the self inductance of the turn P01_C01 at DC	1	0.098	0.05	0.045	0.042	0.041

The strongest drop of self inductance takes place between DC and 0.5 kHz, the factor being 10.3. At 0.5 kHz, the eddy currents induced in the stainless steel jacket, which weaken the outer magnetic field of the turn, are very strong compared to the situation without any eddy currents at DC. The outer inductance decreases with increasing frequency and at 300 kHz, the inner inductance of the conductor is relevant to the calculation of self inductance, only.

The cable-in-conduit design of the PF coil conductor also has a strong influence on the frequency dependence of mutual inductance between the conductors shown in Tab. 4.11. The eddy currents induced in the stainless steel jacket of the conductors of the PF coils weaken mutual inductance between the turns. With increasing frequency, mutual inductance of two turns decreases. It continuous to decrease also at very high frequencies compared to the self inductance of the turn, which reaches a convergence value of inner inductance at a frequency of about 300 kHz.

Tab. 4.11: Mutual inductance between the turns P01_C01 and P01_C02 of the PF 6 coil shown for frequencies used for calculating the resonance frequency of the coil.

Frequency	0 kHz	0.5 kHz	13 kHz	32 kHz	200 kHz	300 kHz
Mutual inductance between turns P01_C01 and P01_C02	1.8e-05 H	3.4e-07 H	2.0e-10 H	7.1e-12 H	3.7e-18 H	1.8e-20 H
Coupling factor between turns P01_C01 and P01_C02	0.76	0.14	2e-04	6e-06	4e-12	2e-14

With increasing distance between the turns, mutual inductance between them decreases. At DC, this effect is smaller than at higher frequencies. For example, mutual inductance of turns P01_C01 and P01_C02 is 1.8e-05 H and between P01_C01 and P01_C27 mutual inductance reaches 5.8e-06 H, the factor between these values being 3.1. Selected mutual inductances between turns of the upper pancake at DC are shown in Tab. 4.12. At a frequency of 300 kHz, mutual inductance between turns P01_C01 and P01_C01 is 1.8e-20 H and between P01_C01 and P01_C27 it is only 4.8e-22 H, the factor between these values being 37. Selected mutual inductances between turns of pancake 01 at 300 kHz are shown in Tab. 4.13.

Tab. 4.12: Selected DC mutual inductances and coupling factors between the turns in the upper pancake of the PF 6 coil.

Turn number	P01_C02	P01_C07	P01_C12	P01_C17	P01_C22	P01_C27
Mutual inductance with turn P01_C01	1.8e-05 H	1.1e-05 H	8.7e-06 H	7.4e-06 H	6.5e-06 H	5.8e-06 H
Coupling factor with turn P01_C01	0.76	0.45	0.36	0.31	0.27	0.24

Tab. 4.13: Selected mutual inductances and coupling factors between the turns in the upper pancake of the PF 6 coil at 200 kHz.

Turn number	P01_C02	P01_C07	P01_C12	P01_C17	P01_C22	P01_C27
Mutual inductance with turn P01_C01	1.8e-20 H	1.9e-23 H	7.5e-24 H	4.6e-24 H	3.1e-24 H	4.8e-22 H
Coupling factor with turn P01_C01	2e-12	2e-17	7e-18	5e-18	3e-18	5e-16

Cumulative inductance of each turn was calculated by adding all mutual inductances of a turn to that of all other turns of the PF 6 coil plus the self inductance of this turn. At DC, the value for cumulative inductance of the turn P01_C01 is 3.3E-03 H, which is very high compared to the self inductance of the same turn of 2.40e-05 H. This high value results from the high mutual inductances between the turns at DC shown above. With increasing frequency, mutual inductance decreases faster than the self inductance shown in Tab. 4.10 and Tab. 4.11. At frequencies higher than 5 kHz, mutual inductances between the turns were neglected, because their values were lower than 1 % of the self inductance and only self inductances of the turns were taken into account. Frequency dependence of the cumulative inductance of the turn P01_C01 is shown in Tab. 4.14.

Tab. 4.14: Frequency dependence of cumulative inductance of the PF 6 coil turn P01_C01.

Frequency	0 kHz	0.5 kHz	13 kHz	32 kHz	200 kHz	300 kHz
Cumulative inductance of turn P01_C01	3.3e-03 H	2.8e-06 H	1.2e-06 H	1.2e-06 H	1.0e-06 H	9.9e-07 H

The total inductance for the PF 6 coil calculated with the detailed FEM model was compared with the inductance of the PF 6 coil calculated with the simplified model of the CS PF coil system. Total inductances at DC were compared, because inductance calculation with the simplified model of the CS PF coil system was made for DC, only. All cumulative inductances of each turn calculated with the detailed FEM model of the PF 6 coil were added to the total inductance of the PF 6 coil. In addition, self inductance of the PF 6 coil was calculated at DC with the formula given in [Gro62] and compared with ITER coil specifications [PAF6Bd]. The calculated total inductances of the PF 6 coil with four calculation methods are summarised in Tab. 4.15.

The relatively small differences show that calculation of total inductance with the formula given in [Gro62] and simplified FEM model of a coil may be used as a good estimation. More detailed calculations of inductances and field analysis of the coil were made with the detailed FEM model of the coil.

Tab. 4.15: Comparison of cumulative total inductance of the PF 6 coil calculated at DC with different methods.

Calculation method	PF 6	Relative difference to detailed FEM model
Detailed FEM model	2.06 H	-
Simplified FEM model	2.02 H	1.9 %
Formula from [Gro62]	2.00 H	2.9 %
Derived from [PAF6Bd]	2.04 H	1.0 %

4.4 FEM Models of Bus Bars for Power Supply

The coil terminals will be connected to the AC/DC converters by two kinds of bus bars. The superconducting bus bars will be used for connections inside the cryostat of ITER. The connection outside of the cryostat will be made by water-cooled bus bars. The resistances, inductances and capacitances of the bus bars for power supplies of the coils were calculated with the FEM program. The calculated values will be used in the network model of the CS PF coil system as lumped elements.

Superconducting Bus Bars

The superconducting bus bars will be used to connect the coil terminals with the Coil Terminal Box (CTB) which will be located outside of the vacuum vessel and the coil cryostat [DDD06i]. The FEM model of the cross-section of a superconducting bus bar is shown in Fig. 4.8.

The modelling of the superconducting bus bars follows the approach that has been used for PF conductor in section 4.2.1. Vacuum was taken to be the material for the cooling tube because of the magnetic properties similar to helium. The inner diameter of the cooling tube and its metallic structure were neglected, because this cooling tube is located inside of the current-leading superconducting cable and has no influence on its inductance and capacitance to ground. The definition of cooling tube as a vacuum creates a current-free space inside the superconducting bus bar with a homogeneous current distribution which is important to the calculation of the inner self inductance of the bus bar.

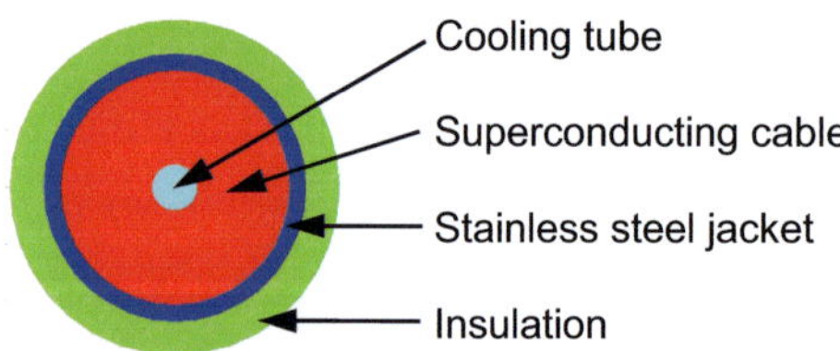

Fig. 4.8:   FEM model of the superconducting bus bar for the power supply of the PF coils.

The cable of the superconducting bus bar consists of 1800 superconducting filaments [DDD06h] and was defined to be a stranded conductor in the FEM model. This means that the current density in the cable is uniform due to the transposed strands of the cable, thus, no skin and proximity effects occur. Copper with a conductivity of 6.4e9 S/m at 4.5 K [DRGAa] was defined to be the material for the FEM model of the superconducting cable, because a specific material has to be used when defining transposed strands in the FEM program.

The stainless steel jacket surrounding superconducting cable was specified to have the conductivity of 1.88e6 S/m at 4.5 K according to [DRGAa]. The insulation of the superconducting bus bars is made of polyimide film-glass impregnated with epoxy resin and has the same dielectric constant as the insulation of the PF coils with $\varepsilon_r = 4$ [DRG1a]. All relevant dimensions of the superconducting bus bars are summarised in Tab. 4.16.

Tab. 4.16: Dimensions of the PF and CS superconducting bus bars derived from [DDD06h].

Outer cooling tube diameter	8 mm
Cable space diameter	41 mm
Stainless steel jacket thickness	3 mm
Outer conductor diameter	47 mm
Insulation thickness	6 mm

The results of the FEM calculations for the superconducting bus bars are summarised in Tab. 4.17. The resistance of the superconducting bus bars calculated by the FEM model was ignored and set to 100 pΩ/m, because the superconducting filaments of the bus bars have no electrical resistivity during normal operation. The value of 100 pΩ/m was needed for the network model, because it does not work without any resistance in the circuit. To calculate the inductance, resistance and capacitance to the ground, the calculated values per metre will be multiplied by the length of the bus bars for each coil of the CS PF coil system.

Tab. 4.17: Results of the FEM calculation for superconducting bus bars.

Inductance	1.40 µH/m
Capacitance to ground	1.08 nF/m

Water-cooled Bus Bars

Water-cooled bus bars are needed to connect the FDU and SNU with the AC/DC converters of the coils. The FEM model of the cross-section of the water-cooled bus bar is shown in Fig. 4.9. The dimensions were taken from [PPSD2]. The bus bar has a height of 106 mm and width of 200 mm. The diameter of the water cooling tubes is 20 mm. Their centre is located 50 mm left and right and 27 mm away from the upper and lower side of the bus bar. The material of the bus bar is aluminium with a conductivity of 3.3e7 S/m at continuous operation temperature of 60°C, according to [DRGAb]. As the cooling tubes are located inside the current-leading bus bar, they have no influence on the outer inductance and capacitance of the bus bar. Vacuum was chosen to be the material for the cooling tubes because of its magnetically and electrically neutral properties.

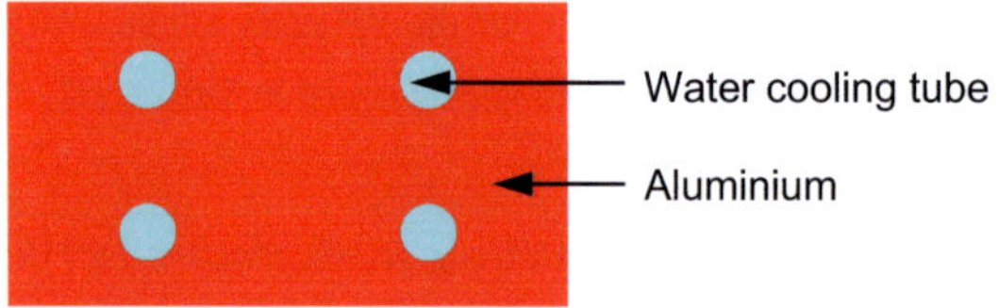

Fig. 4.9:   FEM model of the water-cooled bus bar for power supply of the PF coils.

The results of the FEM calculations are shown in Tab. 4.18. To calculate the inductance, resistance and capacitance to the ground, the FEM values are multiplied by the length of the bus bars for each coil of the system.

Tab. 4.18: Results of the FEM calculation for water-cooled bus bars at DC.

Inductance	1.39 µH/m
Resistance	1.39 µΩ/m
Capacitance to ground	12.6 pF/m
Capacitance to parallel bus bar	34.8 pF/m

Compared to the values specified for the water-cooled bus bars at a resistance of 2.97 mΩ/m in [PPSA1], the resistance calculated with FEM is more than 2000 times smaller. Calculation of the resistance of the bus bars with the height, width and specified conductivity yields 1.43 µΩ/m which is close to the value of 1.39 µΩ/m calculated with FEM. Hence, the resistance value from [PPSA1] is wrong. For further calculations, the calculated resistance of 1.39 µΩ/m was used. Compared to the value of the inductance of the water-cooled bus bars, which was given in [PPSA1] to be 1 µH/m, the calculated value of 1.39 µH/m differs by 40 %. For further calculations, the calculated FEM value for the inductance of 1.39 µH/m was used.

4.5   Assumptions Used in FEM Models

For convenience the assumptions used in FEM models are listed below:

- The transition of the conductors between the pancakes and double pancakes was neglected thus the conductors have rotational symmetry only.

- The cooling tube of 10 mm inner and 12 mm outer diameter with its metallic properties was replaced by a tube with an outer diameter of 12 mm and vacuum as the filling material. This was done, because the cooling tube is placed inside the current-leading superconducting cable and, thus, has no influence on the outer inductance and capacitance of it. The definition of the cooling tube as a vacuum creates a current-free space inside the superconducting cable with a homogeneous current distribution which is important to be calculation of the inner inductance of the conductor.

- The superconducting cable of the PF coils is made of up to 1440 superconducting strands, which means that the influence of skin and proximity effects in the cables is very low. Furthermore, the stranded and twisted design allows for a uniform current density in the whole cross-section of the cable. To consider the cable design in the FEM model of the superconducting cable, the cross-section of the cable, excluding the cross-section of the cooling tube, was defined as a stranded copper conductor with a conductivity of 6.4e9 S/m at 4.5 K. Because of the stranded model, no skin and proximity effects occur in the superconducting cable. The jacket was defined as a solid material that still has a skin and proximity effect.

- In the FEM model, outer corners of the jacket are rectangular, which is in contradiction to the specified rounded corners with 3 mm radius.

- Several kinds of insulation, e. g. turn, layer and ground insulation, were combined to one insulation with a homogeneous permittivity of $\varepsilon_r = 4$ [DRG1a].

- One layer FEM model of the coil was used in FEM calculations with frequencies higher than 5 kHz to manage the memory problems of Maxwell 2D. The magnetic coupling between the turns at these frequencies is lower than 1 % and was neglected.

- The sub-cable has a 0.1 mm thin stainless steel wrap with 50 % coverage to allow for helium penetration [PAC1]. A stainless steel final cable wrap with 0.2 mm total thickness and 50 % overlap protects the cable when it is pulled through the jacket. The superconducting cable and the stainless steel jacket were included in the model as separated and insulated parts. The stainless steel wrap was added to the stainless steel jacket in the detailed FEM models of the PF 3 and PF 6 coils.

- The value for the water-cooled bus bars specified in [PPSA1] for a resistance 2.97 mΩ/m and an inductance of 1 µH/m were not used in the network models. Instead, the FEM values for the resistance of 1.39 µΩ/m and inductance of 1.39 µH/m were used in further calculations.

4.6 Summary of FEM Calculation Results

The FEM calculations yielded inductances of modelled components, which were used as lumped elements for setting up network models. Self and mutual inductances of CS and PF coils were calculated with a simplified FEM model of the CS PF coil system at DC. The frequency dependence of the inductances of the PF 3 and PF 6 coils was analysed with detailed FEM models of each coil.

The PF 3 and PF 6 coils had a strong frequency dependence of the inductance of their conductors, caused by eddy currents induced in the stainless steel jacket of the superconducting cable. The self inductance of one turn decreases with increasing frequency down to 3.3 % of its DC value at a frequency of 300 kHz. Mutual inductances between single turns and between the coils also show a strong frequency dependence caused by eddy currents induced in the stainless steel jackets. The magnetic coupling decreased with increasing frequency and was neglected for calculations at frequencies higher than 5 kHz. Thus, the total cumulative inductance of PF coils decreases down to 0.26 ‰ of its DC value at a frequency of 300 kHz.

The total inductance of PF 3 and PF 6 coils calculated at DC with detailed FEM models was compared with the inductances of both coils calculated at DC with a simplified FEM model of the CS PF coil system. The results are shown in Tab. 4.19.

Tab. 4.19: Comparison of the total inductance of the PF 3 and PF 6 coils at DC calculated with the simplified model of the CS PF coil system and detailed FEM model of the coils.

	PF 3	PF 6
Simplified FEM model	1.83 H	2.02 H
Detailed FEM model	1.94 H	2.06 H
Relative difference	5.6 %	1.9 %

Moreover, the inductances and resistances of superconducting and water-cooled bus bars were calculated at DC to complete the values needed for building up the network model of the power supply circuits of the CS PF coil system.

The FEM values were included in the network models of the coil power supply circuits with simplified CS and PF coils and in the detailed network models of the PF 3 and PF 6 coils as lumped elements. The building of network models and implementation of the elements is described in the next chapter.

5 Network Models

The network program OrCAD PSpice [Orc07] allows to analyse the generated models in the frequency and time domain. This chapter describes the set up of the network model with lumped elements. The inductances were calculated with FEM models of the CS PF coil system at DC and frequency-depended detailed FEM models of PF 3 and PF 6 coils, as described in chapter 4. The network model of the complete CS PF coil system with fast discharge units (FDU), switching network units (SNU) and power supplies (PS) of the coils is shown in section 5.1. The detailed network models of the PF 3 and PF 6 coils with and without instrumentation cables are described in sections 5.2 and 5.3, respectively.

5.1 ITER Coil System

The network model of the CS PF coil system, which is described in this section, was used to calculate the voltage waveforms at the terminals of the PF 3 and PF 6 coils. These calculated voltages were used in detailed network models of the coils as excitation voltages acting on the coil terminals. The voltage at each turn of the PF 3 and PF 6 coils was calculated as system response to these excitations. The calculation results are described in chapter 6. The detailed data of the network model of the CS PF coil system are given in Annex A.2.

The simplified schematic representation of the network model of all CS and PF coil power supply circuits is shown in Fig. 5.1. The circuits of the CS3U, CS2U, CS2L, CS3L, PF 1 and PF 6 coils were single-coil electrical circuits which consist of a current source, switching network unit (SNU), fast discharge unit (FDU) and the coil represented as a single inductance. The coil modules CS1U and CS1L were switched in series in one electrical circuit consisting of one current source, two FDUs, two SNUs and the CS1U and CS1L coil modules represented as single inductances. The PF 2, PF 3, PF 4 and PF 5 coils were switched in parallel in one electrical circuit consisting of four coil power supplies, four FDUs and the PF 2 – PF 5 coils again represented as single inductances. The main, booster and vertical stabilisation converters were combined to one current source in each branch of the PF 2 – PF 5 coils connected in parallel. The self inductances and magnetic couplings of all coils were calculated with the simplified FEM model of the CS PF coil system described in section 4.1 and also considered in the network model. Mutual inductances of the coils were combined in a coupling matrix which was implemented in network calculations. The details of the circuits are shown in the following figures using the power supply circuit of the PF 6 coil as an example.

The PF coils partly reveal different winding directions to generate magnetic fields for plasma control. The winding direction of the PF 2 – PF 5 coils is specified in [PAF6Ba]. The PF 2 and PF 3 coils have the same winding direction. The PF 4 and PF 5 coils also have the same winding direction, but opposite to the PF 2 and PF 3 coils. The winding direction of the other coils, PF 1, PF 6 and all CS coils, is not explicitly specified in the documentation. In [DDDD3] the winding direction of the CS, PF 3 and PF 6 coils is drawn for the upper and lower pancake of the double pancake and also indicated in the side view of these coils. The winding directions of CS, PF 3 and PF 6 coils are the same in these drawings. Consequently, the winding direction of PF 1 was assumed to be the same as that of the PF 3 coil. Thus PF 4 and PF 5 coils only wound oppositely to the other PF and all CS coils.

The network model of a coil circuit is described in detail using the PF 6 coil and its power supply circuit shown in Fig. 5.2 as an example. To show the details of the network model, it was divided into four parts: the power supply (PS), the switching network unit (SNU), the fast discharge unit (FDU) and the simplified network model of the PF 6 coil, which are shown in detail in Fig. 5.3, Fig. 5.4,   Fig. 5.5 and Fig. 5.6, respectively.

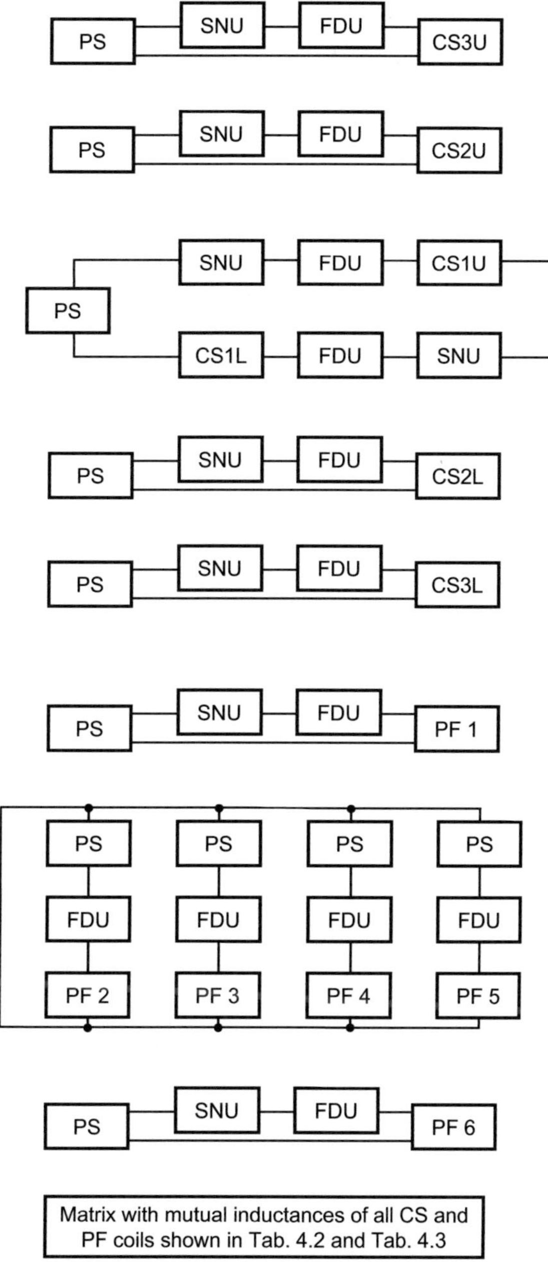

Fig. 5.1: Schematic overview of the network model of the power supply and switching circuits of CS and PF coils, which consists of the coil power supply (PS), switching network unit (SNU), fast discharge unit (FDU) and the simplified models of the coils (e. g. PF 6). Mutual inductances of all CS and PF coils were combined in a coupling matrix in Tab. 4.2 and Tab. 4.3. Fig. 5.2 shows the components of PS, SNU, FDU and PF 6 in detail.

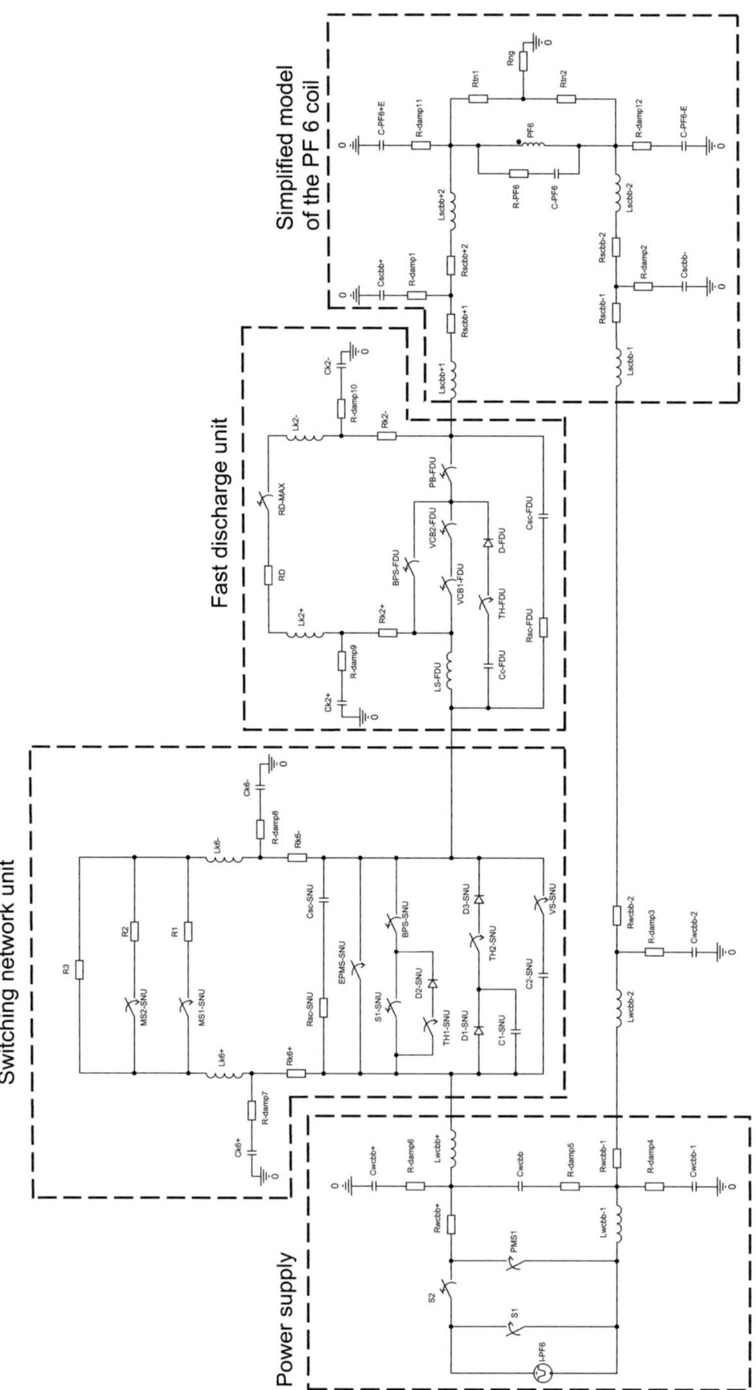

Fig. 5.2:   Network model of the power supply and switching circuit of the PF 6 coil with the parts shown in detail in Fig. 5.3, Fig. 5.4, Fig. 5.5 and Fig. 5.6.

The power supply of the PF 6 coil shown in Fig. 5.3 consists of the current source I-PF6 which supplies the coil circuit with current of up to 45 kA. The free wheeling switch of the power supply S1 was closed before the fast discharge of the coil was started. First, the free wheeling path for the coil current was made by the switch PMS1. The switch S2 separated the power supply from the coil circuit during fast discharge 1 ms after closing PMS1 and S1. In the network model, the switching times of these three switches were set to 1 µs. In reality, the switching times are much slower, but switching times of free wheeling switches can be neglected when calculating the coil voltages.

The power supply was connected to the PF 6 coil and its switching network units via water-cooled bus bars, the FEM model of the bus bars is shown in section 4.4. The values for the water-cooled bus bars were calculated for positive and negative polarity of the connections to the current source. The calculated resistance of 1.39 µΩ/m, inductance of 1.39 µH/m, capacitance to the ground of 13 pF/m and capacitance between the water-cooled bus bars of 35 pF/m were multiplied by the distances between the coil terminal box (CTB) of the coil, the power supply and the switching network unit. These distances were derived from [SLBD1] and [SLBa].

The bus bar with positive polarity was defined to be the connection between the coil power supply and the switching network unit. Its length was estimated from site layouts [SLBD1, SLBa] to be 30 m for the PF 6 coil. The elements of the water-cooled bus bar with positive polarity have the indexes wcbb+PF6. The bus bar with negative polarity was defined to be the connection between the coil terminal box and the coil current source. Its total length was estimated to be 75 m. The first part of the water-cooled bus bars with negative polarity was placed parallel to the positive polarity for about 30 m. This distance is equal to the distance between the switching network unit and power supply of the coils. The other part of the bus bar was placed between the coil terminal box and switching network unit and only has a bus bar in one direction for 31 m and 45 m, respectively, depending on the distance of the coil connection to the current source. The elements of the water-cooled bus bar with negative polarity, which were placed in parallel to the bus bars with positive polarity, had the indexes wcbb-1. The part of the bus bars with negative polarity, which had no parallel bus bar with positive polarity, had the indexes wcbb-2.

The lengths of the water-cooled bus bars for all coils are listed in Tab. 5.1 and Tab. 5.2. The power supplies in the coils PF 2, PF 3, PF 4 and PF 5 connected in parallel were directly connected to fast discharge units, because they had no switching network unit in their electrical circuits. All resistances, inductances and capacitances used for water-cooled bus bars in the network model of the CS PF coil system are given in Tab. A.7 in Annex A.3.

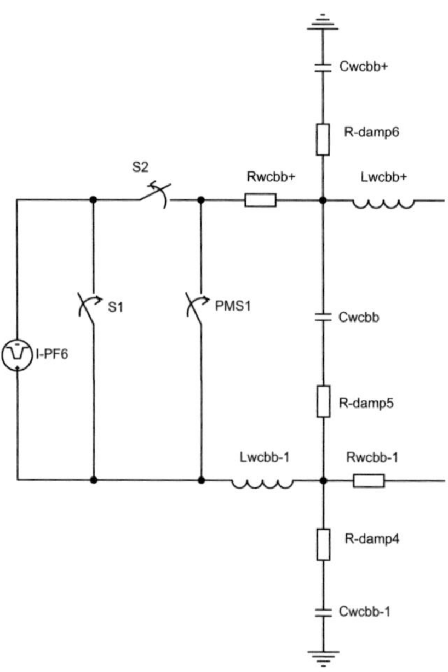

Fig. 5.3:   Network model of the power supply of the PF 6 coil with water-cooled bus bars.

Tab. 5.1:   Length of water-cooled bus bars from the power supply to the switching network unit and coil terminal box for CS coil modules, PF 1 and PF 6 coils [SLBD1, SLBa].

Coil	Length of connection from CTB to power supply	Length of connection from power supply to SNU
CS3U	61 m	30 m
CS2U	61 m	30 m
CS1U	75 m	30 m
CS1L	75 m	30 m
CS2L	75 m	30 m
CS3L	75 m	30 m
PF 1	61 m	30 m
PF 6	61 m	30 m

Tab. 5.2: Length of water-cooled bus bars from the power supply to the fast discharge unit and coil terminal box for PF 2, PF 3, PF 4 and PF 5 coils [SLBD1, SLBa].

Coil	Length of connection from CTB to power supply	Length of connection from power supply to FDU
PF 2	61 m	30 m
PF 3	61 m	30 m
PF 4	75 m	30 m
PF 5	75 m	30 m

The network model of the switching network unit of the PF 6 coil is shown in Fig. 5.4. The values for the elements of the switching network unit were taken from section 2.4 and included in the network model. The switching network resistors R1, R2 and R3 had the same resistance values. The switching network resistors have to supply 10 kV voltages for ignition of the plasma at different currents through the coils [PAF6Bc]. For this purpose, the resistances R1 and R2 can be switched on and off in the switching network unit circuit by the main switches MS1-SNU and MS2-SNU.

Power cables were used for connection between switching network unit and the switching network resistors (SNR). These cables were made of copper with a cross-section of 800 mm^2. The values of the elements for one cable were taken from [Bar05]. The capacitance to ground was 0.6 nF/m, the inductance of one cable was 90 nH/m and the resistance 50 $\mu\Omega$/m. Six parallel cables were used for connecting the switching network unit to the switching network resistors because of the high currents flowing to the resistors during normal operation scenarios. The lengths of the power cable connections between the switching network unit and resistors are given in Tab. 5.3. All capacitances, inductances and resistances of the cables connecting the switching network unit to the resistors are given in Tab. A.8 in Annex A.2.

The snubber circuit which consists of a resistance Rsc-SNU-PF6 of 0.2 Ω and a capacitance Csc-SNU of 0.1 mF limited the high harmonics caused by the switching procedure of the switching network unit. The explosively activated protective make switch EPMS-SNU was used to ensure the free wheeling path for the current and to bridge the switching network unit, if other current switches were disabled.

During normal operation, current flew through the switch S1-SNU and the bypass switch BPS-SNU. The switch TH1-SNU was used to provide the free wheeling current path and to bridge the switching network unit during normal operation. The diode D2-SNU, together with the switch TH1-SNU, was used to simulate the function of a thyristor. The diode D1-SNU was used to limit current flow in one direction. The capacitor C1-SNU of 20 mF was charged up to a voltage of 1 kV. The switch TH2-SNU and the diode D3-SNU simulated the thyristor used to discharge the capacitor C1- SNU and provide for the currentless opening of the bypass switch BPS-SNU. The capacitor C2-SNU = 0.8 mF was charged up to a voltage of 8 kV. The vacuum switch VS-SNU was used to discharge the capacitor C2-SNU and to support the currentless opening of the bypass switch BPS-SNU.

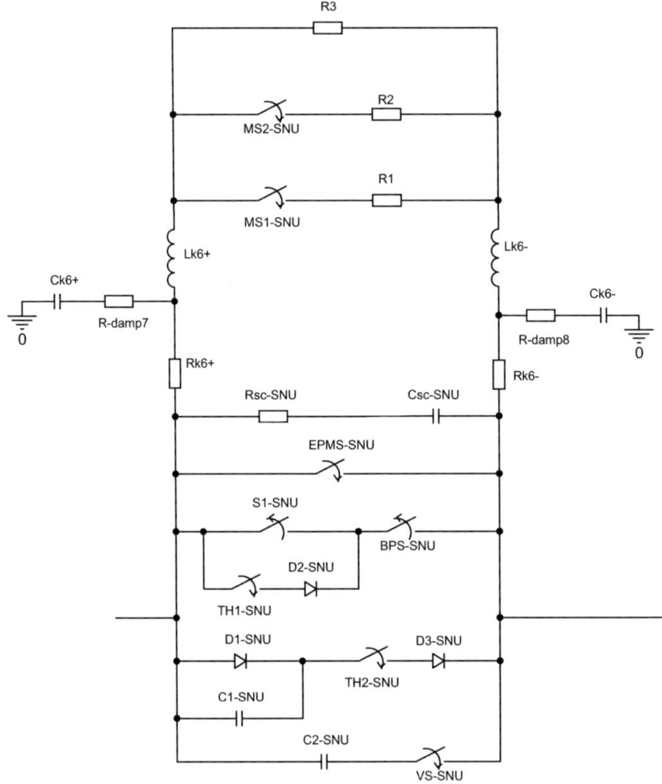

Fig. 5.4:   Network model of the switching network unit (SNU) of the PF 6 coil.

The fast discharge unit was used to commutate the current to the discharge resistor during fast discharge of the coils. The network model of the fast discharge unit of the PF 6 coil is shown in Fig. 5.5. The discharge resistor of the PF 6 coil RD had an initial value of 92 mΩ. During fast discharge, resistance increased due to the heating of the resistor to 216 mΩ. This process was simulated by an additional switch RD-MAX. Its initial resistance was set to 100 pΩ. The resistance of the RD-MAX increased during the closing time of the switch and reached its maximum of 124 mΩ. The closing time of RD-MAX was set to 14 s which corresponded to the current discharge time constant of the PF 6 coil.

The power cables used for connecting the fast discharge unit to the discharge resistor were made of copper with a cross-section of 800 mm^2. The values for one cable were taken from [Bar05]. The capacitance to ground was 0.6 nF/m, the inductance of one cable 90 nH/m and the resistance

50 µΩ/m. For the connection of the fast discharge unit to the discharge resistor, only two parallel cables were used, because the currents to the discharge resistor will flow during the fast discharge process only. The lengths of power cable connections between the fast discharge unit and discharge resistor are listed in Tab. 5.3. All resistances, inductances and capacitances of the cables connecting the fast discharge unit and discharge resistor are given in Tab. A.9 in Annex A.2.

The switch BPS-FDU was used to simulate the bypass switch of the fast discharge unit with the switching times and resistances at the main contacts being the same as those of the real bypass switch. The switches VCB1-FDU and VCB2-FDU were used to simulate the vacuum circuit breaker. Each switch had two resistance levels which were changed during switching time. The vacuum circuit breaker had three resistance levels during current commutation: first if the vacuum circuit breaker is closed with 60 µΩ, second if an electric arc occurs in vacuum circuit breaker by opening of the main contacts at DC with a resistance of 0.8 mΩ and third if the main contacts were opened and no current flew through the vacuum circuit breaker, defined to 100 MΩ. The switch VCB1-FDU switched the resistance from 60 µΩ to 0.8 mΩ during the arcing in the vacuum circuit breaker. The VCB2-FDU increased the resistance of the vacuum circuit breaker to 100 MΩ during the extinction of the arc in the vacuum circuit breaker. The pyrobreaker PB-FDU commutates the current in case the vacuum circuit breaker did not work properly.

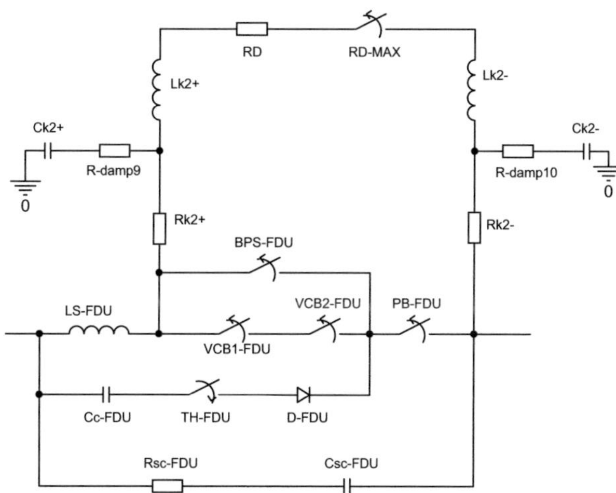

Fig. 5.5: Network model of the fast discharge unit (FDU) of the PF 6 coil.

The circuit with the capacitance Cc-FDU of 1.8 mF, which was charged to a voltage of 7 kV, the switch TH-FDU and the diode D-FDU were used to extinguish the arc in the VCB by discharging Cc-FDU. The switches TH-FDU and D-FDU were used to block the current in reverse direction in the thyristor of the circuit. The switching time of switch TH-FDU was set to

2 µs, according to [Dat1]. The snubber circuit consisting of a resistance Rsc-FDU of 0.15 Ω and a capacitance Csc-FDU of 0.35 mF limited the high harmonics caused by the switching process of the fast discharge unit.

The network model with the simplified PF 6 coil, the superconducting bus bars and symmetrical grounding of the coil is shown in Fig. 5.6. The coil terminal box was connected to the PF 6 coil via superconducting bus bars. The FEM model of the bus bars was shown in section 4.4. To simulate the superconductivity of the filaments, the resistance of the superconducting bus bars was set to 100 pΩ. With the calculated inductance of 1.4 µH/m and capacitance of 1.79 nF/m and with the distances between the coil terminal box and the coil derived from [SLBD1, SLBa], the values for the superconducting bus bars were calculated for positive and negative polarity of the connections of the PF 6 coil, which have the same length of 25 m. The elements of the positive polarity of the superconducting bus bar had the indexes scbb+. The elements of the negative polarity of the water-cooled bus bar had the indexes scbb-. The lengths of the superconducting bus bars between the coil terminal box and the coil terminals are given in Tab. 5.3. All inductances and capacitances of the superconducting bus bars connecting the coil terminal box and the coil terminals are given in Tab. A.10 in Annex A.2.

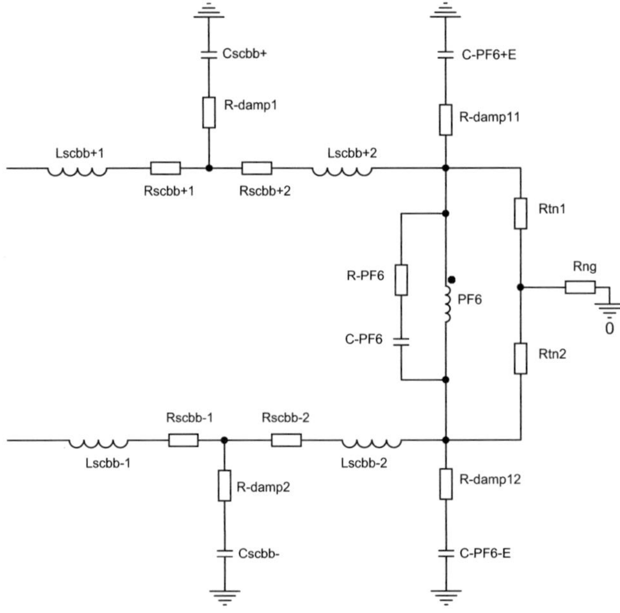

Fig. 5.6:   Network model of the simplified PF 6 coil with superconducting bus bars and symmetrical grounding.

The inductances of the CS and PF coils were calculated using the FEM model of the CS PF coil system and are described in section 4.1. The inductance of the PF 6 coil at DC was calculated to amount 2.01 H. The capacitance of the PF 6 coil was calculated to be 1.78 nF. Total capacitance of each branch as calculated by adding up the capacitances between the layers switched in series. Then, the total capacitance of the whole coil was calculated by adding the branch capacitances which are connected in parallel. The total capacitances calculated for all CS and PF coils are presented in Tab. A.6 in Annex A.3.

Symmetrical grounding of the coil [PAF6Bd] will be ensured by two identical terminal-to-neutral resistances Rtn1 and Rtn2 and the neutral-to-ground resistance Rng with values of 1 kΩ each.

Tab. 5.3: Length of superconducting bus bars and power cables for the connection of switching network unit (SNU) and fast discharge unit (FDU) to switching network resistors (SNR), discharge resistor R_D and coil terminal box (CTB), respectively [SLBD1, SLBa].

Coil	Distance from SNU to SNR with 6 parallel power cables	Distance from SNR to FDU with 6 parallel power cables	Distance from FDU to R_D with 2 parallel power cables	Distance from R_D to CTB with 2 parallel power cables	Distance from CTB to coil with superconducting bus bars
CS3U	36 m	36 m	39 m	39 m	34 m
CS2U	44 m	44 m	47 m	47 m	36 m
CS1U	58 m	58 m	19 m	20 m	40 m
CS1L	55 m	30 m	50 m	80 m	40 m
CS2L	28 m	28 m	33 m	33 m	39 m
CS3L	20 m	20 m	30 m	32 m	37 m
PF 1	10 m	10 m	20 m	20 m	25 m
PF 2	-	-	30 m	30 m	25 m
PF 3	-	-	35 m	35 m	23 m
PF 4	-	-	20 m	40 m	27 m
PF 5	-	-	20 m	40 m	26 m
PF 6	10 m	10 m	25 m	47 m	25 m

The network model of the power supply of the CS and PF coils was used for calculation of the voltages waveforms on the coil terminals for four calculation scenarios, the results of these calculations are shown in section 6.2. The following sections 5.2 and 5.3 shows the detailed network models of PF 3 and PF 6 coil used for calculation of internal voltage distribution.

5.2   Detailed Model of the Poloidal Field 3 Coil

The detailed network models of the PF 3 coil were used to calculate the resonance frequency of the coil in the frequency domain and voltage waveforms between the turns of the PF 3 coil during normal operation, fast discharge and two failure cases in the time domain. The voltage waveforms calculated at the terminals of the PF 3 coil with the network model of the CS PF coil system for four calculation scenarios in the time domain were included as excitations at the coil terminals in the detailed network model of the PF 3 coil. The detailed network model without instrumentation cables will be described in the first section. The influence of the instrumentation cables will be described in the second section. The detailed data of the network models of the PF 3 and PF 6 coils are given in Annex A.3.

Network Model without Instrumentation Cables

The inductances calculated with detailed FEM models of the PF 3 coil shown in section 4.2 were included in the network model as lumped elements. The resistances of NbTi superconductors at 4.5 K could be neglected due to superconductivity. Frequency independence of the resistance of the NbTi superconductor was verified by the experiments with small-sized superconducting NbTi coils described in Annex A.2. However, PSpice does not work correctly without any resistances in electrical circuits. Thus, the resistances of superconducting cables were set to 100 pΩ.

The capacitances of the PF 3 coil between adjacent turns and the outer turns to ground in R-direction were calculated with the formula for cylinder capacitance (5.1). The inner radius (r_i) of the capacitance corresponded with the outer radius of the inner turn. The outer radius (r_o) of the capacitance corresponded with the inner radius of the outer turn. The arrangement of turns and variables is shown in Fig. 5.7 and was used in formula (5.1).

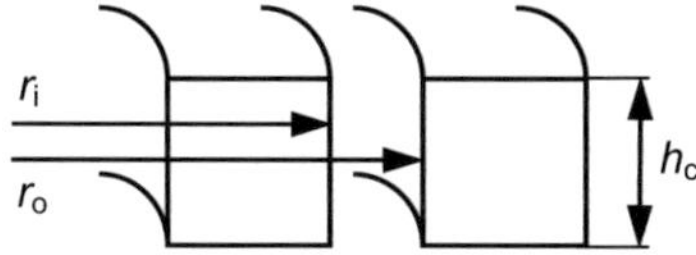

Fig. 5.7:   Capacitance built by two turns in one layer in R-direction.

$$C_{cyl} = 2\pi\varepsilon_0\varepsilon_r \frac{h_c}{\ln(r_o / r_i)}$$

[5.1]

The capacitances of the PF 3 coil between adjacent turns and the outer turns to ground in Z-direction were calculated with the formula for parallel plate capacitance (5.2). The outer radius (r_o) of the capacitance was also the outer radius of both turns. The inner radius (r_i) of the capacitance was also the inner radius of both turns. The arrangement of the turns and variables is shown in Fig. 5.8 and was used in formula (5.2).

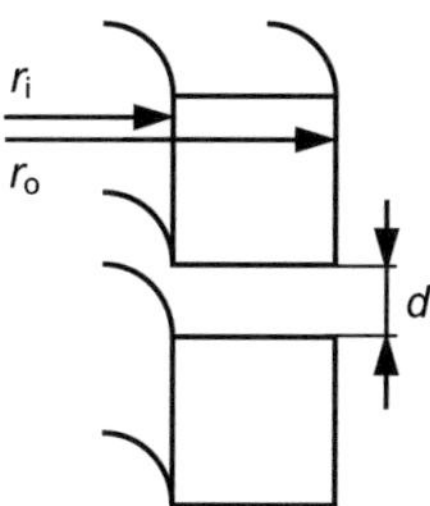

Fig. 5.8:   Capacitance built by two turns in different layers (Z-direction).

$$C_{pp} = \pi\varepsilon_0\varepsilon_r \frac{r_o^2 - r_i^2}{d}$$

[5.2]

The capacitances between the turns of one pancake were calculated between 19.7 nF and 20.7 nF depending on the location of the turns in the coil. Capacitance was lower on the inner side of the pancake, the higher capacitance value was valid to the outer side of the pancake. The capacitances between the turns of two pancakes in one double pancake were calculated from 18 nF to 19.3 nF. The capacitances between the turns of two different double pancakes varied between 14 nF and 15 nF. The capacitances of the outer turns to ground were calculated to range between 9.6 nF and 20.5 nF. The detailed results of the capacitance calculations are given in Annex A.3.

The elements of the network model were denoted as follows: The first letter is the element type, e. g. R for resistance. The number of the pancake is next and is separated from the element designation by an underline, e. g. R_P01. If the element was placed between two pancakes, then the number of the second pancake was added, also separated by another underline, e. g. R_P01_02. The number of the conductor (i. e. turn) follows, also separated by an underline, e. g. R_P01_C01. If the designation was given for an element placed between two conductors, the number of the second conductor was added, again separated by an underline, e. g. R_P01_C01_02. For example, the capacitance between conductors 10 and 11 of the pancake 01 was denoted C_P01_C10_11. The capacitance of conductor 12 between the pancakes 01 and 02 had the designation C_P01_02_C12. The capacitors to ground for the outer conductors had an E placed between the pancake and conductor number, e. g. C_P01_E_C12.

Transformation of the inductances calculated with the FEM program into network elements is shown in Fig. 5.9. At first, the inductances of each turn were calculated with detailed FEM models of the turns shown in Fig. 5.9 part a. The calculated inductances were included in the detailed network model of the turns as lumped elements and divided into two inductances, e. g. the inductance of turn P01_C01 was separated into L_P01_C01/1 and L_P01_C01/2. The calculated capacitances between the turns was set between the two inductances of each turn, e. g. one pin of C_P01_C01_02 were set between L_P01_C01/1 and L_P01_C01/2. Thus, the capacitance C_P01_C01_02 was connected parallel to the second half of inductance of turn P01_C01 (L_P01_C01/2) and the first half inductance of turn P01_02 (L_P01_C02/1) shown in Fig. 5.9 part b. The inductances connected in series were added up to one inductance, with first half consisting of half of the inductance of turn P01_C01 and the other half of half of the inductance of turn P01_C02, e. g. L_P01_C01_02 was the sum of L_P01_C01/2 and L_P01_C02/1 shown in Fig. 5.9 part c. The inductances at the beginning and end of one pancake remained as half elements and were connected to the next pancake or coil terminal, e. g. L_P01_C01/1 and L_P01_C03/2 shown in Fig. 5.9 part c.

The detailed network model of the PF 3 coil was built up using the calculated elements. The overview of the detailed network model of the PF 3 coil is shown in Fig. 5.10. The zoomed right upper part of the network model is shown in Fig. 5.11. The coil terminals are located on the right side of the PF 3 network model. The numbers of the conductors were given from the left to the right, the connection of the voltage source being placed on the winding with the greatest number for conductor P01_C12. The number of the turn was placed at the bias point between two inductances, e.g. P01_C12 shown in Fig. 5.11. This point also was the connection for the capacitor to the adjacent turn and for the capacitors to the ground. The values for the turns at the beginning and the end of a pancake remained with the values divided by two, e. g. R_P01_C12/2 and L_P01_C12/2 shown in Fig. 5.11. All resistances and inductances connected in series were combined to one resistance and inductance, respectively. Thus, the number of resistances and inductances was reduced by a factor of nearly two.

The detailed network model of the PF 3 coil was used for calculations in the frequency and time domains. The resonance frequency of the coil was calculated in the frequency domain. The voltage waveforms between the turns were calculated in the time domain.

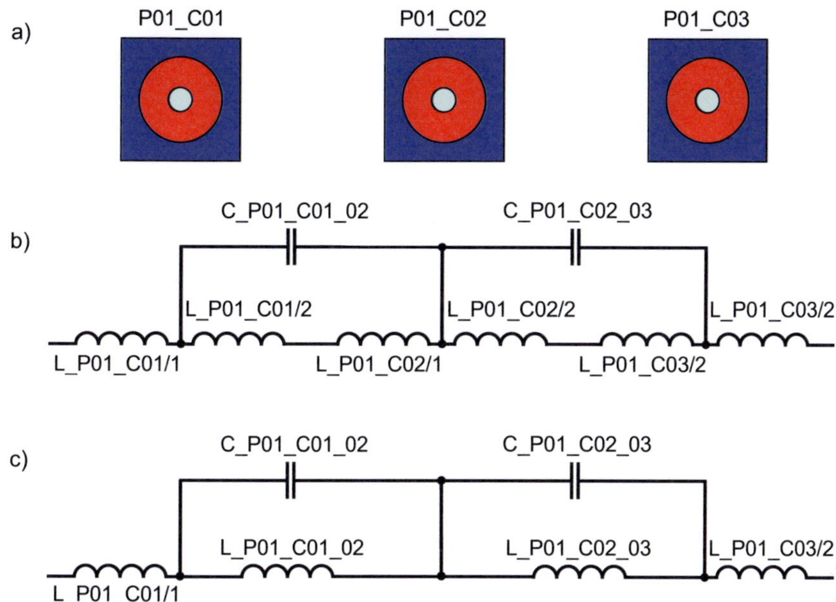

Fig. 5.9: Transformation of FEM-calculated inductances into network model elements consisting of three parts.

Part a: FEM model of three conductors in one pancake.

Part b: Network with calculated inductances of each conductor separated into two inductances, e. g. L_P01_C01/1 and L_P01_C01/2.

Part c: Network with cumulated inductances to reduce the number of elements, e. g. L_P01_C01/2 and L_P01_C02/1 are summed to L_P01_C01_02.

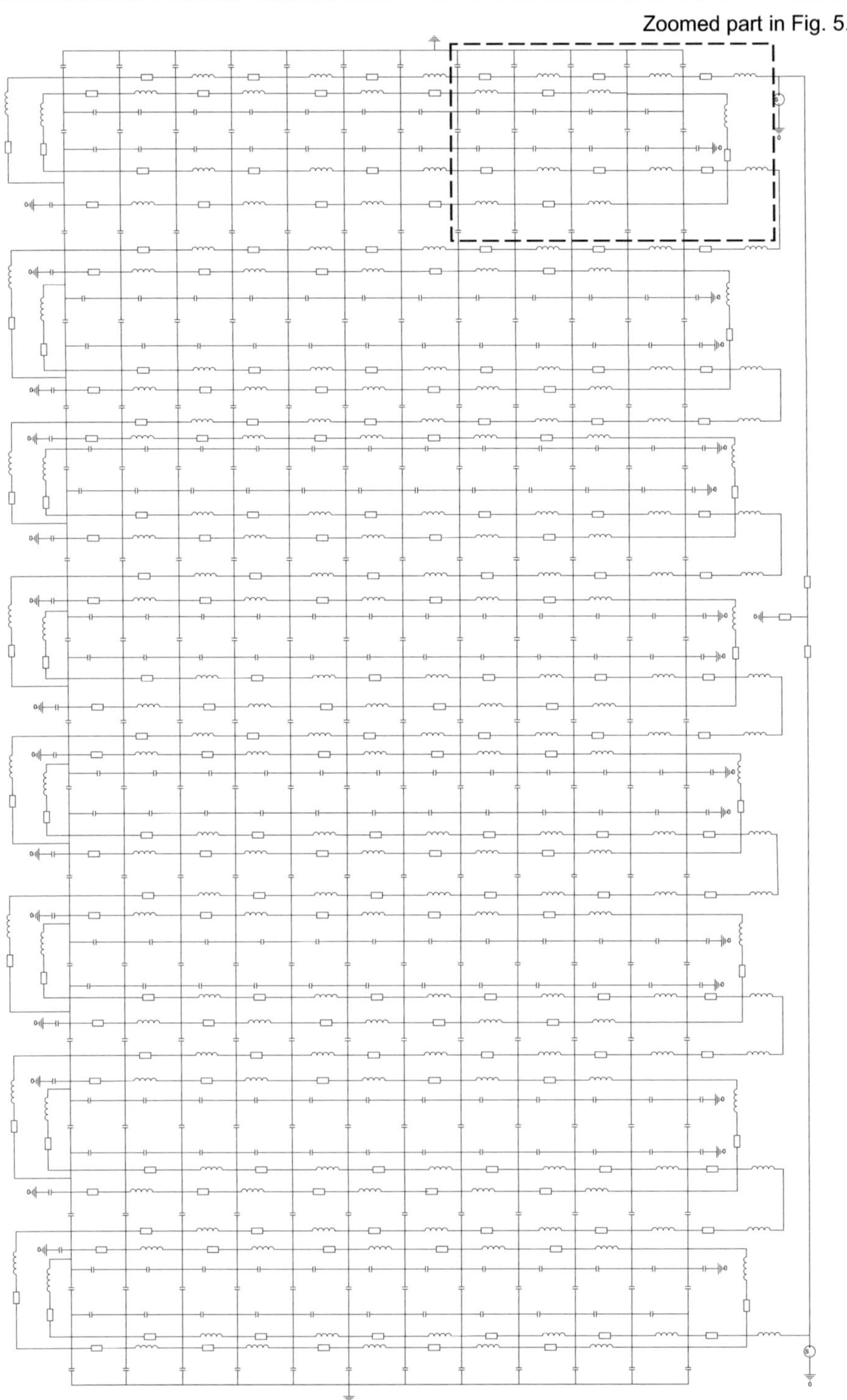

Fig. 5.10: Detailed network model of the PF 3 coil with the zoomed part shown in detail in Fig. 5.11.

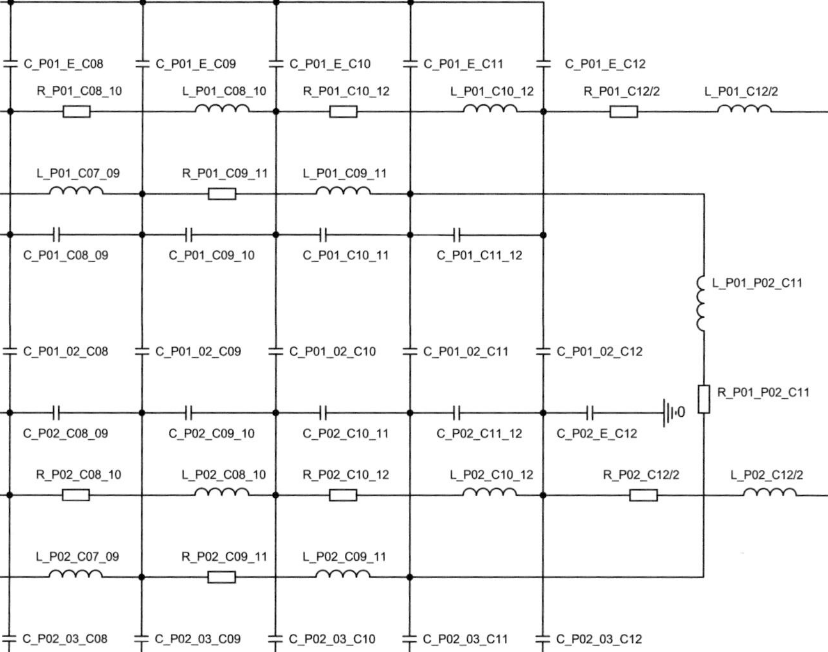

Fig. 5.11: Zoomed right upper part of the detailed network model of the PF 3 coil.

Network Model with Instrumentation Cables

The influence of the instrumentation cables on the resonance frequency of the coils was analysed by consideration of the instrumentation cables in the detailed network model. The instrumentation cables will be placed at the joints between the double pancakes, but the type of cables is not exactly defined in [CDA3]. Calculations were made with the same instrumentation cable type [Bau98] which was used for the ITER Toroidal Field Model Coil (TFMC) [Ulb05]. The capacitance of these cables was calculated to be 437 pF/m. The cables were included in the network model as additional capacitances on the double pancake joints. The current limiting resistor of 50 kΩ [CDA3] which is planned to be used at the beginning of the ITER instrumentation cable was neglected. Using of the 50 kΩ current limiting resistor leads to the same frequency behaviour as without instrumentation cables. The designation IC was used for the capacitances of the instrumentation cables. The lengths of the instrumentation cables were derived from [CDA3, DDDD4]. The lengths of the cables and their capacitances are listed in Tab. 5.4.

Tab. 5.4:   Length and capacitance values for instrumentation cables of the PF 3 coil.

	Length of instrumentation cable	Capacitance of instrumentation cable
C_IC_E_P01	53 m	23.0 nF
C_IC_E_P02_03	45 m	19.6 nF
C_IC_E_P04_05	62 m	27.0 nF
C_IC_E_P06_07	45 m	19.6 nF
C_IC_E_P08_09	32 m	14.0 nF
C_IC_E_P10_11	46 m	20.0 nF
C_IC_E_P12_13	62 m	27.0 nF
C_IC_E_P14_15	46 m	20.0 nF
C_IC_E_P16	60 m	26.0 nF

5.3   Detailed Model of the Poloidal Field 6 Coil

The detailed network models of the PF 6 coil were used to calculate the resonance frequency of the coil in the frequency domain and voltage waveforms between the conductors of the PF 6 coil during normal operation, fast discharge and two failure cases in the time domain. The voltage waveforms calculated at the terminals of the PF 6 coil with the network model of the CS PF coil system for four calculations scenarios in the time domain were included as excitations in the detailed network model of the PF 6 coil. The detailed network model without instrumentation cables will be described in the first section. The influence of the instrumentation cables will be described in the second section. The detailed data of the network model are given in Annex A.3.

Network Model without Instrumentation Cables

The inductances calculated with detailed FEM models of the PF 6 coil shown in section 4.3 were included in the network model as lumped elements. The resistances of NbTi superconductors at 4.5 K are very low and frequency-independent, which was shown by the experiments with small-sized superconducting NbTi coils described in chapter A.2. On the other hand, PSpice does not work correctly without any resistances in electrical circuits. For this reason, the resistances of the superconducting cables were set to 100 pΩ.

The calculation of the values for the capacitances of the PF 6 coil between the conductors and between the outer conductors and ground was made with formula for cylindrical capacitance (5.1) and formula for parallel plate capacitance (5.2) with the procedure which was described in section 5.2.

The capacitances between the turns of one pancake were calculated between 6.1 nF and 8.7 nF, depending on the location of the turns in the coil. The lower capacitance value was found on the inner side of the pancake, the higher capacitance value on the outer side of the pancake. The capacitances between the turns of two pancakes in one double pancake varied from 5.8 nF to 8.5 nF. The capacitances between the turns of two different double pancakes were calculated to range between 5.1 nF and 7.4 nF. The capacitances of the outer turns to the ground varied between 3.1 nF and 8.9 nF. The detailed results of the capacitance calculations are given in Annex A.3.

The detailed network model of the PF 6 coil was built up with the calculated elements. The overview of the detailed network model of the PF 6 coil is given in Fig. 5.12. The zoomed right upper part of the network model is shown in Fig. 5.13. The elements were designated in the same way as those of the PF 3 coil and described in section 5.2.

Network Model with Instrumentation Cables

The influence of the instrumentation cables on the resonance frequency was analysed by consideration of the instrumentation cables in the detailed network model. The instrumentation cables will be placed at the joints between the double pancakes, but the type of cables is not exactly defined in [CDA3]. Hence, calculation was done using the same instrumentation cable type [Bau98] as that used for the ITER Toroidal Field Model Coil (TFMC) [Ulb05]. The capacitance of these cables was calculated to be 437 pF/m. The cables were included in the network model as additional capacitances on the double pancake joints. The current limiting resistor of 50 kΩ [CDA3] which is planned to be used at the beginning of the ITER instrumentation cable was neglected. Using of the 50 kΩ current limiting resistor leads to the same frequency behaviour as without instrumentation cables. The designation IC was used for the capacitances of the instrumentation cables. The lengths of the instrumentation cables were derived from [CDA3, DDDD4]. The lengths of the cables and their capacitances are listed in Tab. 5.5.

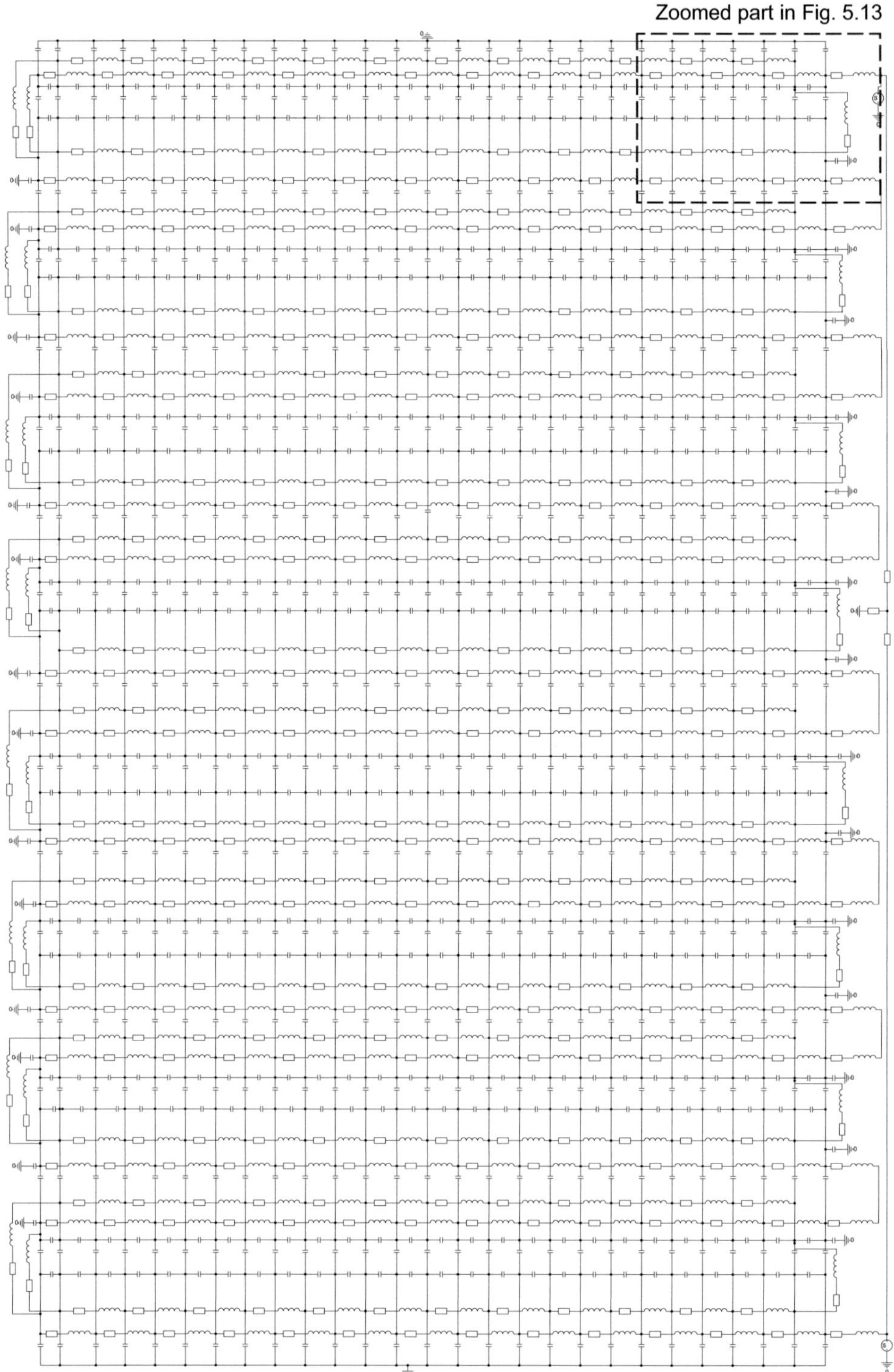

Fig. 5.12: Detailed network model of the PF 6 coil with the zoomed part shown detail in Fig. 5.13.

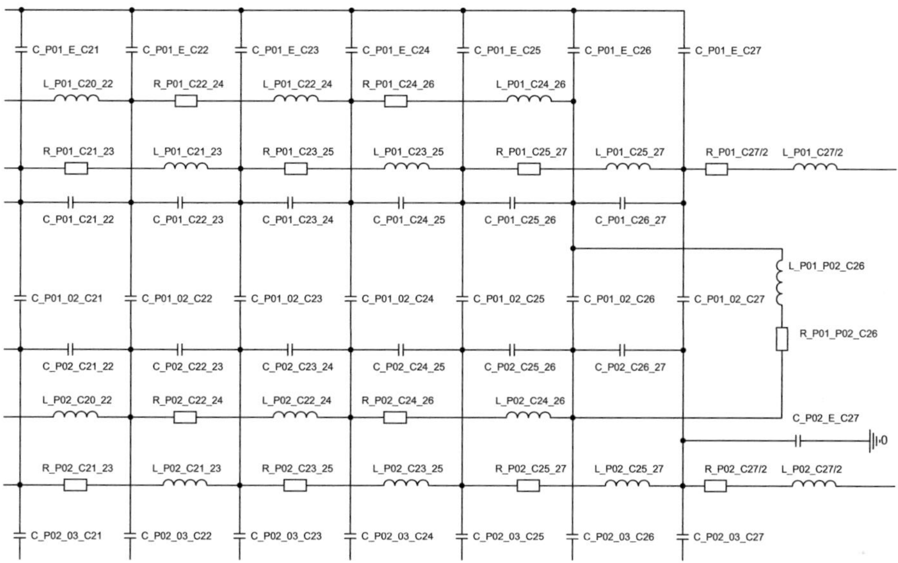

Fig. 5.13: Zoomed right upper part of the detailed network model of the PF 6 coil.

Tab. 5.5: Length and capacitance values for instrumentation cables of the PF 6 coil.

	Length of instrumentation cable	Capacitance of instrumentation cable
C_IC_E_P01	44 m	20.0 nF
C_IC_E_P02_03	34 m	15.0 nF
C_IC_E_P04_05	38 m	16.5 nF
C_IC_E_P06_07	41 m	18.0 nF
C_IC_E_P08_09	46 m	20.0 nF
C_IC_E_P10_11	46 m	20.0 nF
C_IC_E_P12_13	41 m	18.0 nF
C_IC_E_P14_15	38 m	16.5 nF
C_IC_E_P16	46 m	20.0 nF

5.4 Assumptions Used in the Network Models

- In the detailed network models of the PF 3 and PF 6 coils only one RLC circuit was used for each turn of the coils.
- All mutual inductances and self inductance of one turn in the detailed network models of the PF 3 and PF 6 coils were added up to one cumulative inductance of this turn. At low frequencies, the capacitances between the turns were neglected due to their high capacitive reactance. At high frequencies, the mutual inductances between the turns were neglected due to their low values compared to the self inductance of each turn.
- The connection of the instrumentation cables at the joints between the pancakes and double pancakes in the detailed network models of the PF 3 and PF 6 coils was considered to be an additional capacitance at the joints.
- The vertical stabilisation AC/DC converter (VSC) was neglected during the network simulation of the CS PF coil system, because it will be by-passed during fast discharge.
- In the PF 2 – PF 5 coil network, the main and booster converters in each branch were combined to one converter which is represented by a current source.
- In the CS1L - CS1U coil network, the two main converters of each coil were combined to one main converter in the network, because both converters are connected in series.
- The winding direction of the PF 1 coil was assumed to be the same as that of PF 2, PF 3, PF 6 and CS coils. The PF 4 and PF 5 coils were wound in opposite direction [DDDD3].

6 Calculation Results

The resonance frequency in the frequency domain and voltage waveforms in the time domain were calculated with the detailed network models of the PF 3 and PF 6 coils. First, the resonance frequency of both coils was calculated to give a benchmark regarding the transient behaviour of the coils. The calculation process and results in the frequency domain are described in section 6.1. The calculations in the time domain were started with voltage waveforms at the terminals of the PF 3 and PF 6 coils calculated with the network model of the CS PF coil system for four calculation scenarios described in detail in section 6.2. The calculated voltage waveforms were included in the detailed network models of the PF 3 and PF 6 coils as voltage excitations on coil terminals. The internal voltage distribution was analysed with three detailed network models of each coil for different frequencies and four calculation scenarios. The internal voltage distribution, maximum voltages and voltage waveforms on ground, layer and turn insulation are shown in detail in section 6.3.

6.1 Resonance Frequency

The resonance frequency of the PF 3 and PF 6 coils was calculated in the frequency domain and gives a first benchmark regarding the transient behaviour of these coils. The influence of instrumentation cables and unsymmetrical grounding on resonance frequencies of the coils was also analysed in the frequency domain. The calculation of resonance frequency includes an iterative loop due to the frequency dependence of the inductances of both coils. The calculation strategy shown in Fig. 6.1 will be explained using the PF 3 coil as an example. All the calculation steps were also used for the calculation of the resonance frequency of the PF 6 coil.

At the beginning of the calculations, the detailed FEM model of the coil was set up with specified geometries and materials, as was described in section 2.2. Because of the unknown resonance frequency of the coil, the starting point for the calculations was set to 0 Hz. The frequency-dependent inductances were calculated with the FEM model of the coil for the frequency of 0 Hz. The frequency-independent capacitances were calculated using formulas for the cylindrical and parallel plate capacitor, as was shown in section 5.2. The superconducting cables have no resistance during all calculation scenarios shown by experiments with two NbTi superconducting coils in Annex A.2. On the other hand, the network program causes a drift of calculated voltage without any resistance in the circuit. Therefore, the resistances of the superconducting cables were defined to be 100 pΩ as a compromise. The calculated inductances, capacitances and specified resistances were included into the network model as lumped elements.

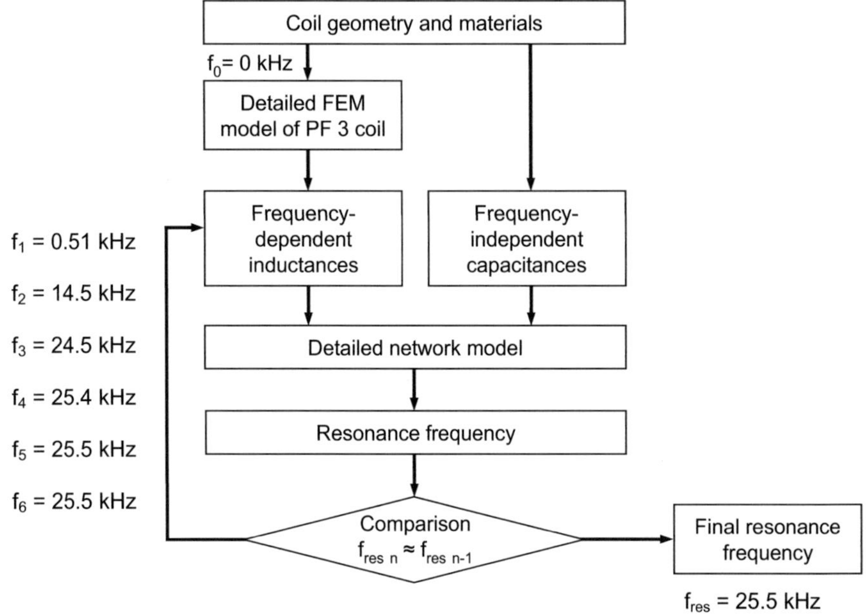

Fig. 6.1:   Strategy of calculation of the resonance frequency using the PF 3 coil as an example.

The calculations in the frequency domain were made by exciting the detailed network model of the PF 3 coil with a voltage of 1 V. The response to excitation was calculated for all frequencies in a defined frequency range, which is shown by the example of the transfer function of voltage calculated on turn P08_C02 in Fig. 6.2. The first calculated resonance frequency was taken into account, because it is the natural frequency of the coil and it is closer to high-harmonic frequencies of AC/DC converters. High voltage peaks were calculated at frequencies higher than 1 kHz shown in Fig. 6.2 part a. The closer view in the frequency range lower than 0.9 kHz shows a voltage peak at 0.51 kHz with a value of 1.07 V in Fig. 6.2 part b. Consequently, the first resonance frequency of the first loop was calculated to be 0.51 kHz.

The comparison between the calculated frequency of the first loop of 0.51 kHz and the frequency of 0 Hz, at which the FEM model of the coil was set up, reveals a relatively big difference. Hence, a second calculation loop was started. The new values for the frequency-dependent inductances were calculated at 0.51 kHz. The values were included in the network model as lumped elements and the resonance frequency of the second loop was calculated to be 14.5 kHz. The relative difference between the resonance frequency of the first loop of 0.51 kHz and that of the second loop of 14.5 kHz was still big, thus the third loop of calculation was started. The inductances were calculated with the FEM model for the frequency of 14.5 kHz and included in the detailed network model of the PF 3 coil. The resonance frequency of the third loop was calculated to be 24.5 kHz.

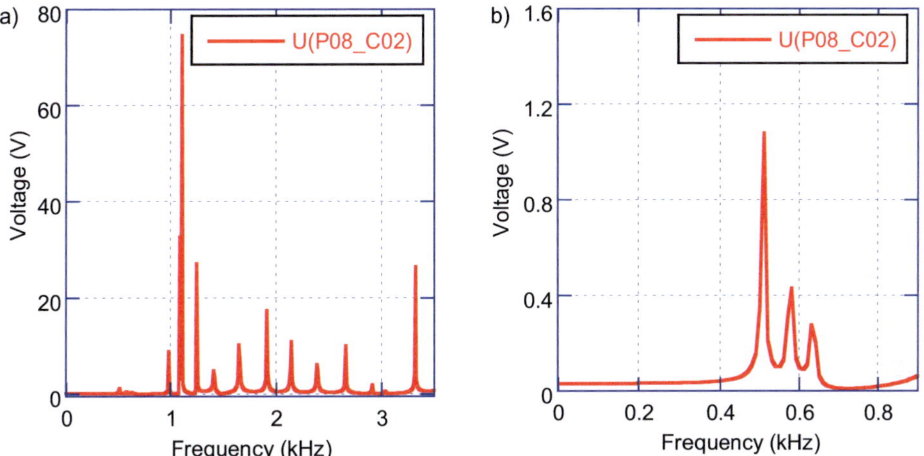

Fig. 6.2: Transfer function of the voltage on the turn P08_C02 for an excitation of 1 V at the coil terminals of the PF 3 detailed DC network model. Part a: in a frequency range up to 3.5 kHz. Part b: in a frequency range up to 0.9 kHz.

The difference of 0.1 % between the resonance frequency calculated with the network model and the frequency at which the FEM model was generated was defined to the breakout criterion of the calculation loop. This criterion was reached after the sixth calculation loop and so, the final resonance frequency of 25.5 kHz was calculated for the PF 3 coil. The transfer function of the PF 3 coil calculated with the detailed network model at 25.5 kHz is shown in Fig. 6.3. The resonance frequencies higher than 25.5 kHz were neglected for further calculations and only the first resonance frequency was taken into account.

To verify that the resonance frequency converges to one value of 25.5 kHz only, the calculations were made with the start frequency of 200 kHz. The calculations strategy was the same as in the case of the start frequency of 0 Hz. The break-out criterion was the resonance frequency of 25.5 kHz. This criterion was fulfilled after the third calculation loop. The results of calculations of the resonance frequency of the PF 3 coil with symmetrical grounding are shown in Tab. 6.1.

Additionally, the influence of the instrumentation cables on the resonance frequency of the PF 3 coil was analysed. The network model of the PF 3 coil with instrumentation cables and symmetrical grounding, which was described in section 5.2, was used for these calculations. The instrumentation cables were represented by capacitors without current limiting resistor of 50 kΩ which is planned be used at the beginning of ITER instrumentation cables [CDA3]. After the fifth calculation loop, the resonance frequency of the PF 3 coil with instrumentation cables was calculated to 24.4 kHz. Comparison of the results between the calculations with and without instrumentation cables shows that the instrumentation cables have an influence on the resonance frequency of the coils and displace it to lower frequencies.

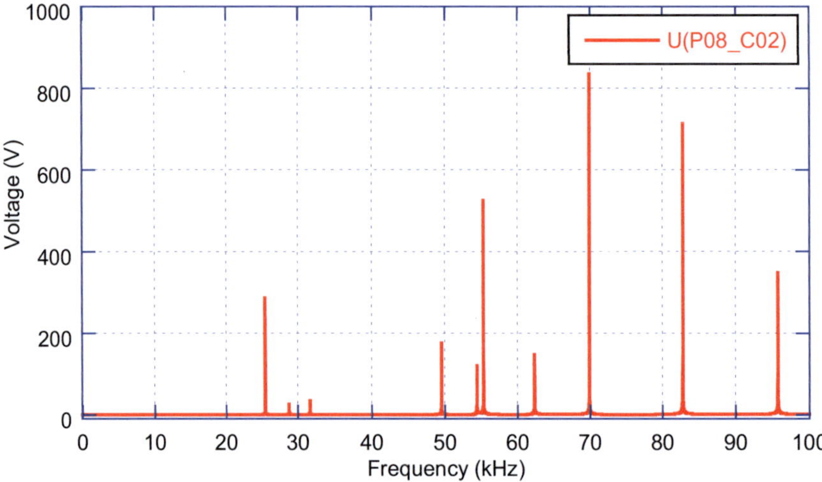

Fig. 6.3:   Transfer function of the voltage on the turn P08_C02 for an excitation of 1 V at the coil terminals of the PF 3 detailed 25.5 kHz network model.

Tab. 6.1:   Calculated resonance frequencies of the PF 3 coil with symmetrical grounding.

	Without instrumentation cables at start frequency 0 Hz	Without instrumentation cables at start frequency 200 kHz	With instrumentation cables at start frequency 0 Hz
1st loop	0.51 kHz	27.8 kHz	0.49 kHz
2nd loop	14.5 kHz	25.6 kHz	14.0 kHz
3rd loop	24.4 kHz	25.5 kHz	23.5 kHz
4th loop	25.4 kHz		24.3 kHz
5th loop	25.5 kHz		24.4 kHz
6th loop	25.5 kHz		

The calculations of the resonance frequency for the PF 6 coil were made with the same strategy as used for the PF 3 coil. The results of the calculations with a start frequency of 0 Hz and

200 kHz, both without instrumentation cables, and the calculations with instrumentation cables at a start frequency of 0 Hz are shown in Tab. 6.2. The resonance frequency of the PF 6 coil without instrumentation cables was calculated to be 32 kHz. The resonance frequency of the PF 6 coil with instrumentation cables was calculated to be 30.7 kHz.

Tab. 6.2: Calculated resonance frequencies of the PF 6 coil with symmetrical grounding.

	Without instrumentation cables at start frequency 0 Hz	Without instrumentation cables at start frequency 200 kHz	With instrumentation cables at start frequency 0 Hz
1st loop	0.55 kHz	34.2 kHz	0.58 kHz
2nd loop	18.7 kHz	32.1 kHz	18.1 kHz
3rd loop	30.8 kHz	32.0 kHz	29.6 kHz
4th loop	31.9 kHz		30.6 kHz
5th loop	32.0 kHz		30.7 kHz
6th loop	32.0 kHz		30.7 kHz

Additionally, the influence of unsymmetrical grounding on the resonance frequency was analysed [Win09], e. g. in case of an earth fault on one coil terminal. The results of the calculation of the resonance frequency with unsymmetrical grounding for the PF 3 coil are summarised in Tab. 6.3. The resonance frequency of the PF 3 coil without instrumentation cables and with symmetrical grounding was calculated to be 25.5 kHz. With unsymmetrical grounding, the resonance frequency is displaced to the lower value of 20.7 kHz. The instrumentation cables at unsymmetrical grounding displace the resonance frequency of the PF 3 coil to much a lower value of 17.0 kHz.

Tab. 6.3: Calculated resonance frequencies of the PF 3 coil with unsymmetrical grounding.

	Without instrumentation cables at start frequency 0 Hz	Without instrumentation cables at start frequency 200 kHz	With instrumentation cables at start frequency 0 Hz
1st loop	0.42 kHz	22.8 kHz	0.35 kHz
2nd loop	11.9 kHz	20.8 kHz	9.50 kHz
3rd loop	19.9 kHz	20.7 kHz	16.3 kHz
4th loop	20.6 kHz		17.0 kHz
5th loop	20.7 kHz		17.0 kHz
6th loop	20.7 kHz		

The resonance frequencies of the PF 6 coil with unsymmetrical grounding are summarised in Tab. 6.4. Unsymmetrical grounding also displaces the resonance frequency of the PF 6 coil to a lower value of 28.6 kHz compared to the resonance frequency of 32.0 kHz with symmetrical grounding of the PF 6 coil. The instrumentation cables have a strong influence on the resonance frequency of the PF 6 coil with unsymmetrical grounding and displace it to 22.9 kHz.

Tab. 6.4: Calculated resonance frequencies of the PF 6 coil with unsymmetrical grounding.

	Without instrumentation cables at start frequency 0 Hz	Without instrumentation cables at start frequency 200 kHz	With instrumentation cables at start frequency 0 Hz
1st loop	0.50 kHz	30.8 kHz	0.40 kHz
2nd loop	16.8 kHz	28.7 kHz	12.9 kHz
3rd loop	27.8 kHz	28.6 kHz	22.1 kHz
4th loop	28.6 kHz		22.9 kHz
5th loop	28.6 kHz		22.9 kHz

Using of the current limiting resistance of 50 kΩ at the beginning of the ITER instrumentation cables leads to the same behaviour as without instrumentation cables, thus the calculations of internal voltage distribution were made with detailed network models of PF 3 and PF 6 coils without instrumentation cables.

6.2   Voltage Waveforms at the Coil Terminals

The voltage waveforms at the terminals of the PF 3 and PF 6 coils in the time domain were calculated with the network model of the CS PF coil system, described in detail in section 5.1. The calculated voltage waveforms at the coil terminals will be included in the detailed network models of the PF 3 and PF 6 coils as excitations for calculation of internal voltage distribution within both coils. The voltage waveforms at the coil terminals of the PF 3 and PF 6 coils were calculated for four calculation scenarios:

- Reference scenario describing the current behaviour of CS and PF coils during normal operation.
- Fast discharge of coils starting, e. g. in case of a quench of one of the superconducting coils.
- Failure case 1, in which the coils were driven at medium voltages and rated currents and an earth fault occurred on negative coil terminal during fast discharge of the coils.
- Failure case 2, in which the coils were driven at rated voltages and medium currents and an earth fault occurred on negative coil terminal during fast discharge of the coils.

The relevant current behaviour of the PF 3 and PF 6 coils, the calculated voltage waveforms at the coil terminals and Fast Fourier Transformation (FFT) of voltage waveforms at the coil terminals during the four calculation scenarios are described in following sections.

6.2.1   Reference Scenario

The reference scenario was defined by plasma physicists to be the current behaviour of all CS and PF coils during the operation of ITER. The reference scenario consists of several operation periods needed for confinement, ignition and control of plasma and induction of a plasma current of 15 MA in a defined section within the vacuum vessel. The current behaviour of the PF 3 and PF 6 coils during the reference scenario with a total duration of 1800 s is shown in Fig. 6.4.

The CS coils are fully energised and the PF coils are preloaded with the specified currents before the start of the reference scenario. The reference scenario begins with the start of discharge (SOD) of the coils at 0 s. The start of plasma (SOP) begins at 0.85 s. The initiation of plasma is the first part of plasma operation and is finished at 1.35 s. At 11.5 s, the X point formation (XPF) begins, which is the transition of limiter plasma to diverter plasma. Start of flat top (SOF) begins at about 70 s when the maximum plasma current is reached and additional heating is applied. Start of burn (SOB) begins at about 90 s and is the time when the rated thermonuclear power generation starts, e. g. 500 MW at a plasma current of 15 MA. End of burn (EOB) is defined to be at about 500 s when plasma cooling is started and the ramp-down of coil currents begins. End of cooling (EOC) begins at about 600 s when plasma cooling is finished and faster ramp-down of coil current starts. End of plasma (EOP) is at about 700 s when the plasma current is decayed to 0 MA. At 975 s, the dwell state between the pulse cycles begins, in which none of the CS and PF coils is energised. The reload of the coils begins at 1490 s and ends at 1790 s. Then they are ready for a new pulse cycle. At 1800 s, the new cycle of the reference scenario begins with SOD.

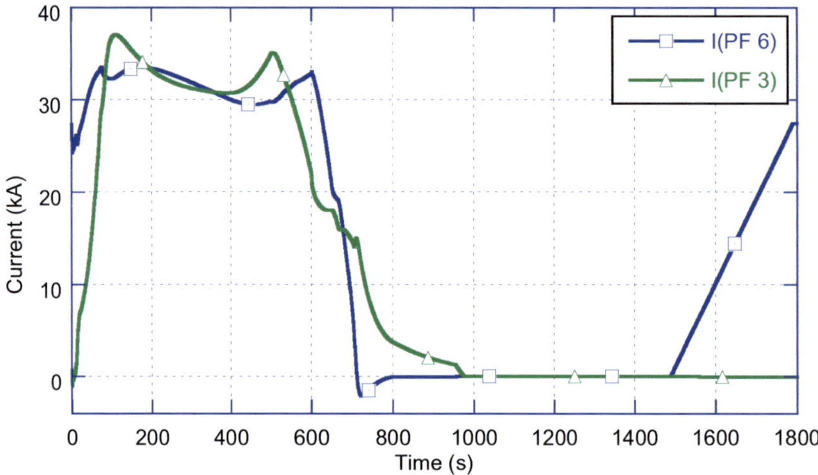

Fig. 6.4:   Current behaviour of the PF 3 and PF 6 coils during the reference scenario specified for 15 MA plasma current operation [Mit09].

The AC/DC converters of the coil power supplies and switching network units (SNU) will supply the voltages required by the PF 3 and PF 6 coils for specified current behaviours during the reference scenario. The voltage waveforms at the positive and negative terminals of the PF 3 coil are shown in Fig. 6.5, the voltage waveforms at the positive and negative terminals of the PF 6 coil are shown in Fig. 6.6, for both coils during the whole reference scenario.

The voltage waveforms of both coils show a similar behaviour during the whole reference scenario. The voltages at the positive and negative terminals reach their highest values at the beginning of the reference scenario with voltages of up to nearly 15 kV at the positive terminal of the PF 3 coil. In other parts of the reference scenario, the terminal voltages do not exceed a voltage level of about 2 kV. The current behaviour of both coils shown in Fig. 6.4 reaches its maximum values at about 100 s of the reference scenario. Despite of this, the change of current is relatively low during the whole reference scenario. Consequently, the maximum voltage at the coil terminals occurs during highest change of current at the beginning of the reference scenario.

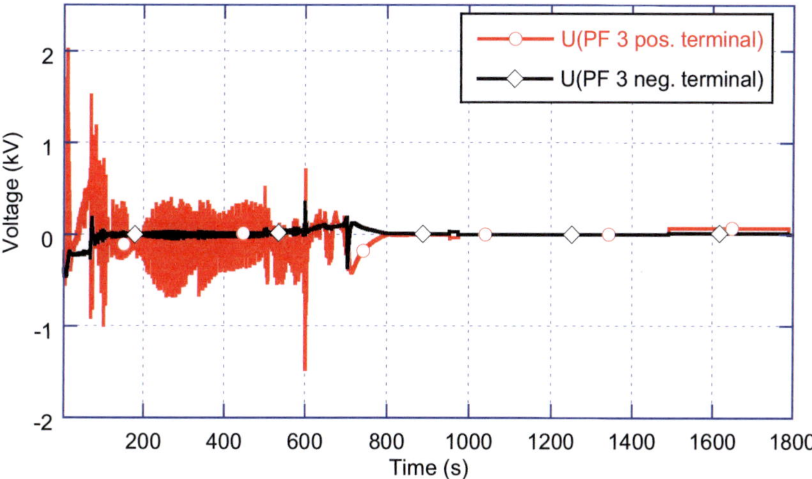

Fig. 6.5:   Calculated voltage waveforms at the positive and negative terminals of the PF 3 coil during the reference scenario without first 1.4 s.. Maximum time of sampling steps is 1 ms.

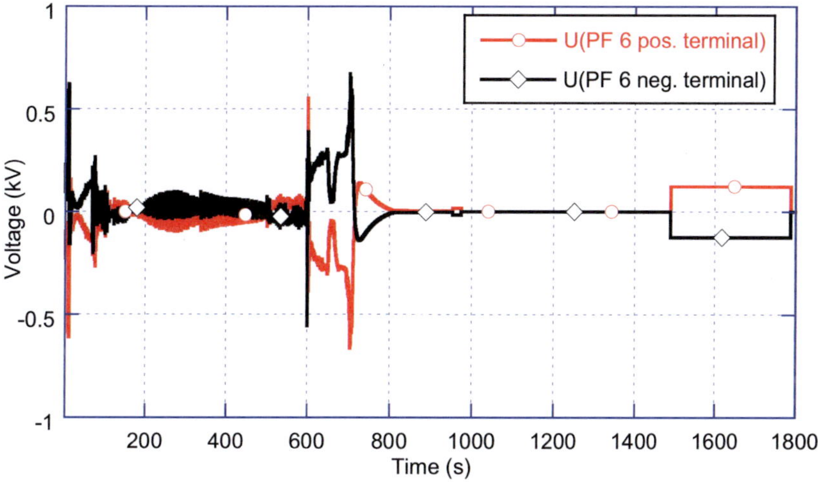

Fig. 6.6:   Calculated voltage waveforms at the positive and negative terminals of the PF 6 coil during the reference scenario without first 1.4 s. Maximum time of sampling steps is 1 ms.

The current behaviour and voltage waveforms at the positive and negative terminals of the PF 3 coil during the first 1.4 s of the reference scenario are shown in Fig. 6.7. The absolute values of the current were relatively low ±1 kA, but the maximum change of the current up to 7 kA/s was calculated during first 1.4 s of the reference scenario. The voltage level at the coil terminals was changed to ensure the derivative of current needed for initiation of plasma. Consequently, a particular voltage level was maintained for several ms to continue the change of currents through the coil.

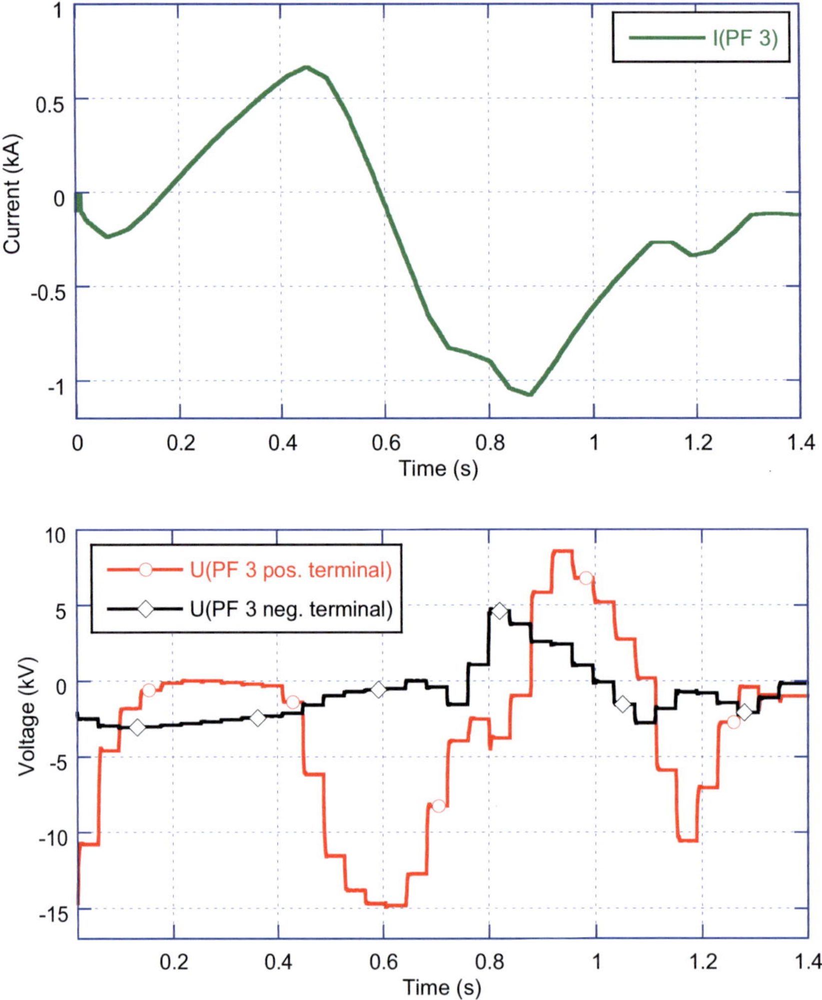

Fig. 6.7:   Specified current behaviour of the PF 3 coil during the first 1.4 s of the reference scenario in the upper part [Mit09] and calculated voltage waveforms at the positive and negative terminals of the PF 3 coil during the first 1.4 s of the reference scenario in the lower part. Maximum time of sampling steps is 1 ms.

The positive and negative terminals of the coils were grounded symmetrically to half the voltage to ground compared to the voltage between the terminals, as it was shown in Fig. 2.10. The voltage waveforms at the terminals of the PF 3 coil during the reference scenario were distributed asymmetrically due to the specific design of its power supply circuit, where the PF 2, PF 3, PF 4 and PF 5 coils were connected parallel and had own main and booster converters in the branches of each coil.

The transitions between the voltage levels were accomplished after a rise or fall time of 2.2 ms, which corresponds to the rise time of sine voltage at about 55 Hz. The rise and fall times were calculated using the 10 % to 90 % values. The voltages at both terminals are sectionally constant voltages in the main parts of the reference scenario. The Fast Fourier Transformations (FFT) of the voltage waveforms at the terminals of the PF 3 coil during the first 1.4 s of the reference scenario are shown in Fig. 6.8. The FFT shows high voltages in the frequency range lower than 5 Hz, which means that the excitation acting on the PF 3 coil terminals during the reference scenario mainly consists of voltage excitations of low frequency.

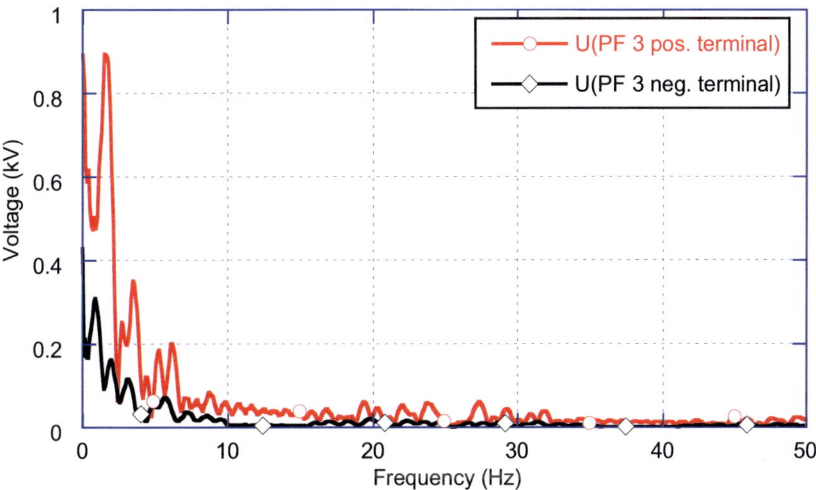

Fig. 6.8:   Fast Fourier Transformation (FFT) of voltage waveforms at the positive and negative terminals of the PF 3 coil during the first 1.4 s of the reference scenario.

The current behaviour and voltage waveforms at the positive and negative terminals of the PF 6 coil during the first 1.4 s of the reference scenario are shown in Fig. 6.9. The derivative of current was highest during the first 1.4 s of the whole reference scenario. Hence, the voltages at the terminals of PF 6 coil were also highest during the first 1.4 s of reference scenario. During the initiation of plasma, the voltage levels at the coil terminals were changed to ensure the change of current. Consequently, particular voltage level was maintained for several ms to continue the derivative of currents through the coil.

The positive and negative terminals of the coils are grounded symmetrically to reduce the voltage to ground to half compared to the voltage between the terminals. Hence, the voltage waveforms at the terminals of the PF 6 coil are distributed symmetrically due to the design of the single coil circuit for driving the PF 6 coil by the own main converter. Additional voltage needed for initiation of plasma was delivered by the switching network unit (SNU).

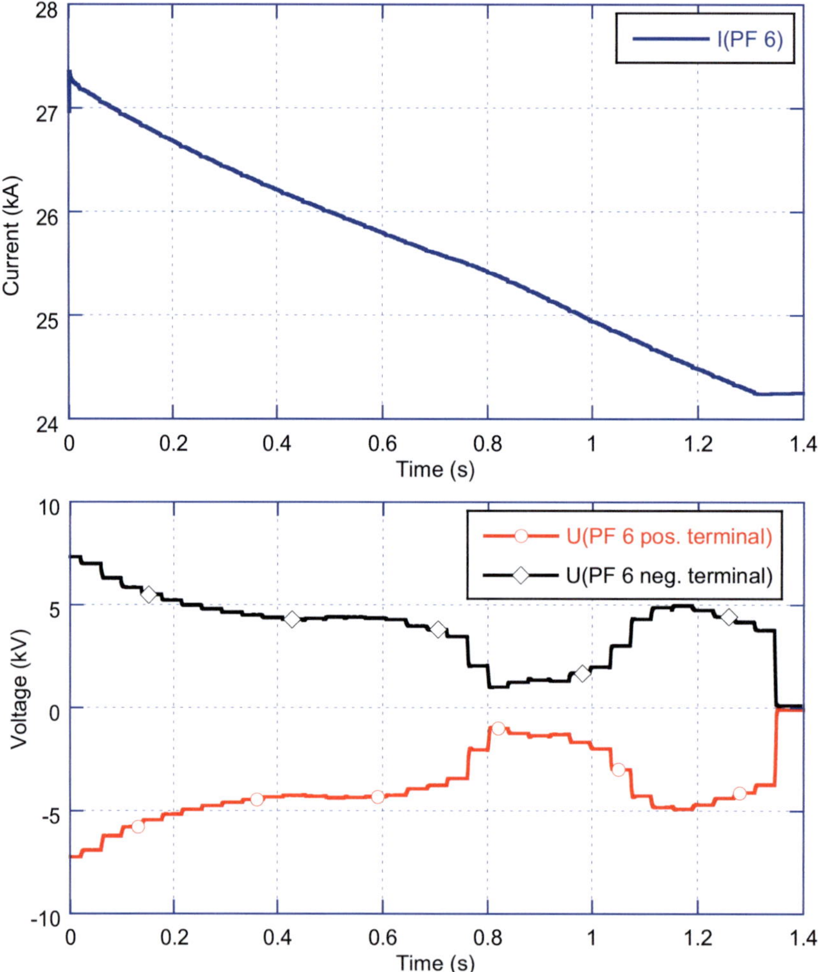

Fig. 6.9: Specified current behaviour of the PF 6 coil during the first 1.4 s of the reference scenario in the upper part [Mit09] and calculated voltage waveforms at the positive and negative terminals of the PF 6 coil during the first 1.4 s of the reference scenario in the lower part. Maximum time of sampling steps is 1 ms.

The transitions between the voltage levels were accomplished after a rise time of 4 ms, which corresponds to the rise time of sine voltage at about 30 Hz. The voltages at both terminals are sectionally constant voltages in the main parts of the reference scenario. The Fast Fourier Transformations (FFT) of the voltage waveforms at the terminals of the PF 6 coil during the first 1.4 s of the reference scenario are shown in Fig. 6.10. The FFT shows high voltages in the frequency range lower than 3 Hz, which means that the excitation acting on the PF 6 coil terminals during the reference scenario mainly consists of voltage excitations of low frequency. In contrast to the PF 3 coil, the FFT of voltage waveforms at both terminals of the PF 6 coil show the same behaviour due to symmetrical grounding and the single coil power supply circuit of the PF 6 coil.

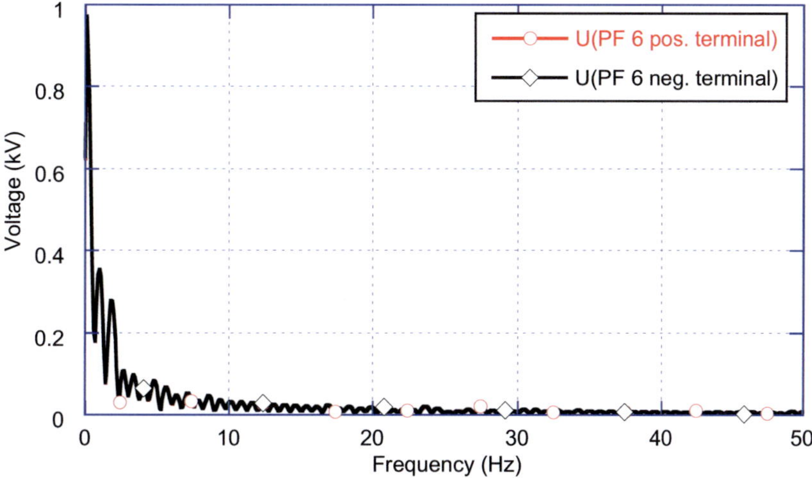

Fig. 6.10: FFT of voltage waveforms at the positive and negative terminals of the PF 6 coil during the first 1.4 s of the reference scenario.

6.2.2 Fast Discharge

Fast discharge of all superconducting coils will start, if a failure appears in one of the CS and PF coils or in the components of the coil power supply circuits. The currents flowing through the coils have to be reduced with an equivalent discharge time constant of 18 s for PF coils and 11.5 s for the CS coil to prevent the superconducting coils from being damaged. All coils of the CS PF coil system will be discharged simultaneously [DDD06m], even if the failure occurs in one of these coils only because of the magnetic coupling in the coil system. The fast discharge units

(FDU) of each coil consist of the same elements except for discharge resistor. Its value depends on the total inductance of the coils to reach the equivalent discharge time constant during fast discharge. The components of the fast discharge unit are shown in detail in Fig. 2.11.

Due to high DC currents flowing through the coils, current commutation from the bypass switch (BPS) to the discharge resistor R_D will be accomplished by a vacuum circuit breaker (VCB) and discharge of counterpulse capacitor C_C. The current behaviour through the components of the fast discharge unit and the PF coil is shown schematically in Fig. 6.11. The fast discharge procedure is divided into four periods with different time scales to explain the function of each element in detail.

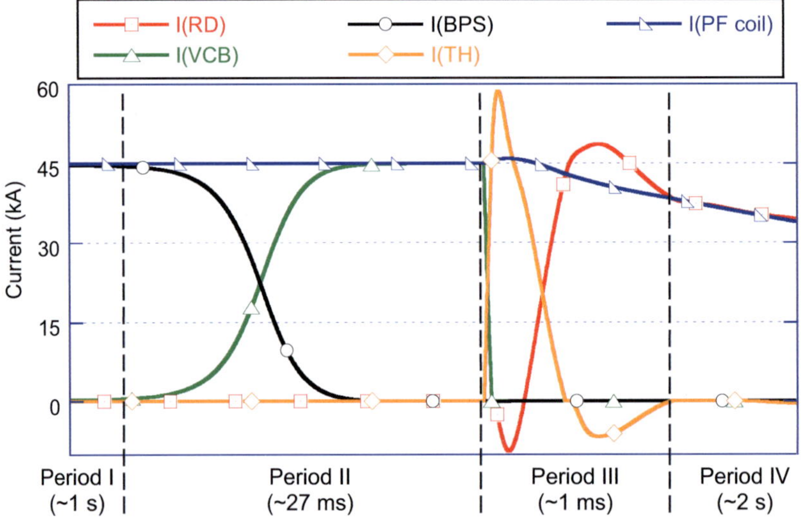

Fig. 6.11: Schematic representation of the current behaviour of the discharge resistor (RD), vacuum circuit breaker (VCB), bypass switch (BPS), ignition thyristor (TH) and PF coil, divided into four periods with different time scales.
Period I: Normal operation before fast discharge.
Period II: Commutation of current from bypass switch to vacuum circuit breaker.
Period III: Ignition of thyristor, discharge of counterpulse capacitor and commutation of current from vacuum circuit breaker to discharge resistor.
Period IV: End of current commutation to discharge resistor and discharge of PF coil. Discharge resistor and PF coil have the same current amplitudes.

In period I, shown for 1 s as an example, the coils will be driven with their current behaviours specified in the reference scenario during normal operation. In period II at the beginning of fast discharge, a free wheeling path for coil current will be created and the power supply of the coils will be disconnected from the coil circuit by switches, which are not components of the fast discharge unit. Then, the bypass switch (BPS) will be opened and the coil current commutates to the parallel vacuum circuit breaker (VCB). After commutation to the vacuum circuit breaker (VCB) is completed, the VCB will be opened and an arc will occur inside the VCB for about 7 ms [Lib08]. Commutation of coil current from bypass switch (BPS) to vacuum circuit breaker lasts about 27 ms. In period III which lasts about 1 ms, the capacitor C_C will be discharged by ignition of thyristor Th and extinguish the arc in the vacuum circuit breaker. The capacitor C_C of 1.8 mF will have been charged to a voltage of 7 kV [PPSA1a] and store enough energy for extinction of the arc in the vacuum circuit breaker. During the discharge of the capacitor C_C, the coil current through the vacuum circuit breaker will commutate to discharge resistor R_D of the coil. The current commutation procedure is completed when the current through discharge resistor R_D reaches the same value as the coil current. In period IV, shown for the first 2 s, the decrease of current begins with the current discharge time constant of 14 s for the PF coils.

The discharge of counterpulse capacitor C_C preloaded with 7 kV for commutation of current from the vacuum circuit breaker to discharge resistor R_D causes a fast increase of voltage at the coil terminals. The voltage waveforms at the terminals of the PF 3 coil during fast discharge are shown in Fig. 6.12. The voltages at the coil terminals during the commutation of current from the vacuum circuit breaker to the bypass switch are relatively low. Therefore, the voltage waveforms are shown from 27 ms when the discharge of the counterpulse capacitor C_C starts by ignition of thyristor Th. The voltages at the both terminals are in steady-state conditions after 1 ms when the current through the discharge resistor R_D reaches the value of the coil current. The voltage between the terminals in steady-state conditions was verified by multiplication of $R_D = 0.08\ \Omega$ by the current through the coils of 45 kA, equalling 3.60 kV. The voltage between the terminals of the PF 3 coil in the network model, in steady-state conditions, 28 ms after the beginning of fast discharge was calculated to be 3.64 kV.

Highest voltages during fast discharge were calculated at the beginning when the counterpulse capacitor C_C is discharged for commutation of current to discharge resistor R_D. The oscillations on the coil terminals caused by the discharge of the preloaded capacitor C_C reach the maximum voltage of -5.3 kV at the negative terminal of the PF 3 coil. The rise time of this voltage is 3.5 μs, which is about 1000 times faster than the rise time of the maximum voltage calculated during the reference scenario. The rise time was calculated with the 10 % to 90 % principle. The maximum voltage at the positive terminal was calculated to be 4.6 kV with a rise time of 3.6 μs. The lower part of Fig. 6.12 shows oscillations of voltage waveforms with equal time periods at the positive and negative terminals of the PF 3 coil. On the other hand, the voltage waveforms at both terminals have a DC off-set voltage, which causes the replacement of the waveforms by positive and negative values, respectively.

The Fast Fourier Transformation (FFT) of voltage waveforms at the positive and negative terminals of the PF 3 coil are shown in Fig. 6.13. Both curves show a similar behaviour in the frequency range up to 50 kHz. In frequency range between 50 kHz and 100 kHz, there are slight differences in the amplitude of the harmonics with high parts at about 90 kHz, which corresponds to the time period of oscillations of about 11.6 µs in the time domain shown in the lower part of Fig. 6.12. The voltage oscillations on both terminals were depended on the inductances and capacitances of the superconducting and water-cooled bus bars. In the frequency range higher than 100 kHz, the FFT of both voltage waveforms shows a similar behaviour despite the peak at about 265 kHz, where the waveform of the positive terminal has voltages with high harmonics parts.

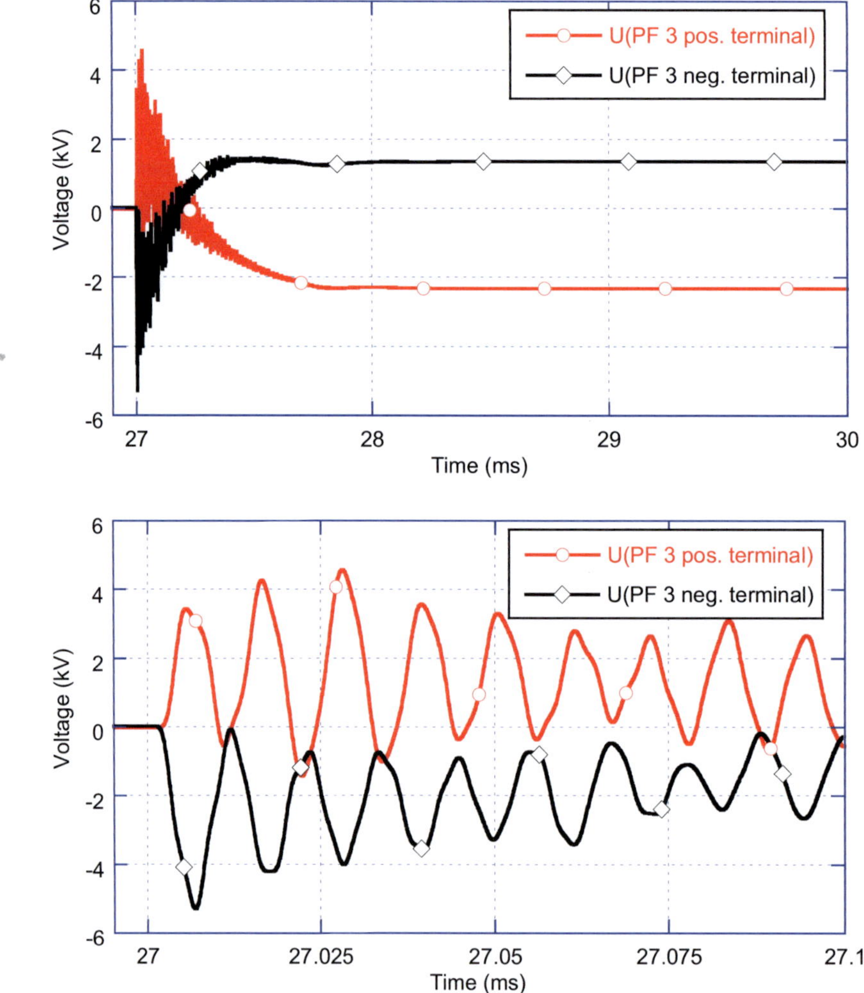

Fig. 6.12: Calculated voltage waveforms at the positive and negative terminals of the PF 3 coil during fast discharge in the upper part on a time scale between 26.9 ms and 30.0 ms and in the lower part on a time scale between 26.99 ms and 27.10 ms. Maximum time of sampling steps is 0.2 µs.

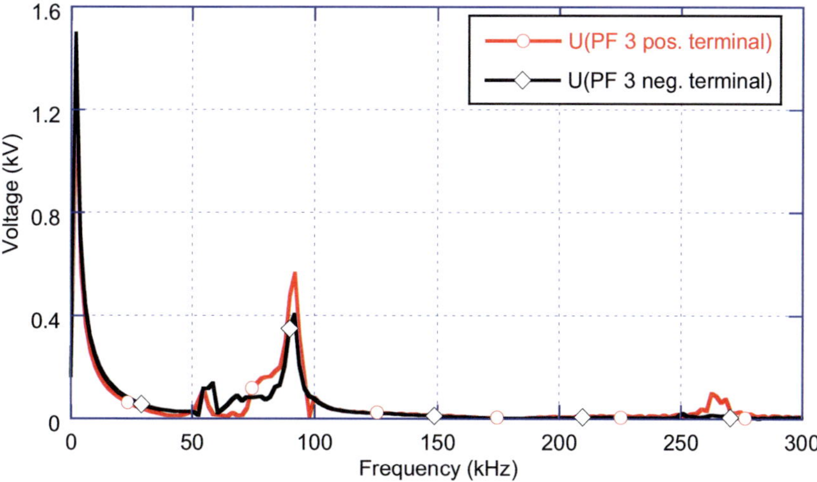

Fig. 6.13: FFT of voltage waveforms at the positive and negative terminals of the PF 3 coil during fast discharge on the time scale between 26.99 ms and 27.10 ms.

The calculated voltage waveforms at the positive and negative terminals of the PF 6 coil during fast discharge are shown in Fig. 6.14. Maximum voltage at the negative terminal of the PF 6 coil was calculated to be -5.0 kV due to voltage oscillations caused by the discharge of the preloaded counterpulse capacitor. The rise time of the maximum voltage is 2.7 µs and comparable with the amplitude and rise time of the maximum voltage at the negative terminal of the PF 3 coil. The maximum voltage at the positive terminal of the PF 6 coil was calculated to be 3.8 kV with a rise time of 1.4 µs. Due to the different lengths of the water cooled bus bars to the terminals of the PF 6 coil and resulting different inductances and capacitances of the bus bars, the oscillations at the positive terminal are faster by a factor of nearly two than the oscillations at the negative terminal.

The FFT of the voltages at the negative and positive terminals of the PF 6 coil shown in Fig. 6.15 confirms the different parts of the harmonics of the terminal voltages for frequencies higher than 70 kHz. The voltage waveform of the positive terminal has high parts at about 190 kHz. The voltage at the negative terminal has high parts at about 95 kHz. The time periods of voltage oscillations at both terminals were calculated to be 5.0 µs and 9.5 µs, respectively, corresponding to the time periods of voltages at the positive and negative terminals of the PF 6 coil in the time domain shown in the lower part of Fig. 6.14. High parts at frequencies lower than 10 kHz were calculated due to the DC voltage at the coil terminals in steady-state conditions after the commutation of coil current to the discharge resistor.

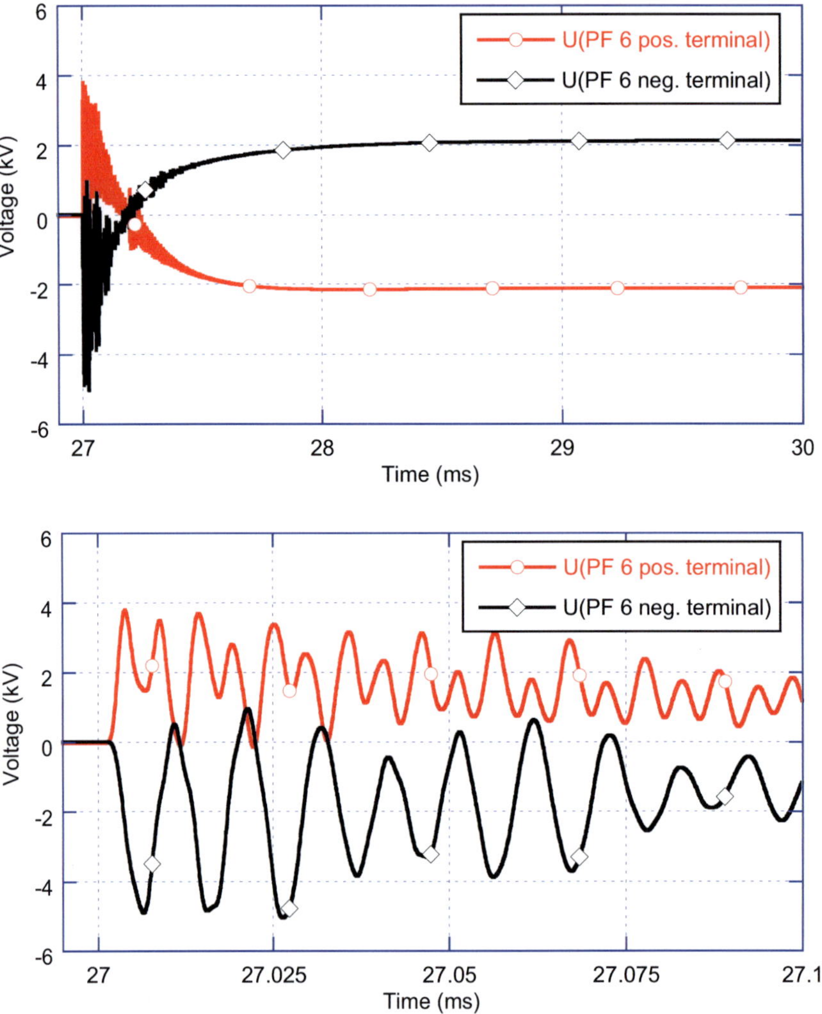

Fig. 6.14: Calculated voltage waveforms at the positive and negative terminals of the PF 6 coil during fast discharge in the upper part on the time scale between 26.9 ms and 30.0 ms and in the lower part on the time scale between 26.99 ms and 27.10 ms. Maximum time of sampling steps is 0.2 μs.

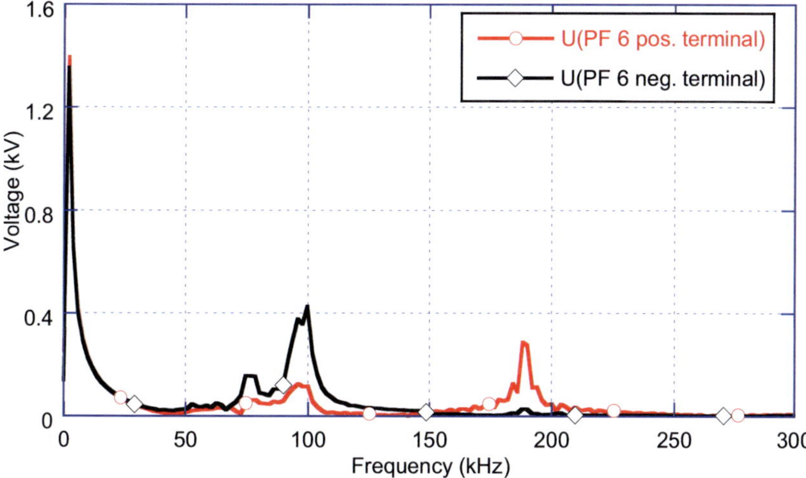

Fig. 6.15: FFT of voltage waveforms at the positive and negative terminals of the PF 6 coil during fast discharge on the time scale between 26.99 ms and 27.10 ms.

6.2.3 Failure Case 1

Failures may occur during the operation of ITER due to its complex construction and large number of components. Two failure cases with different coil currents and terminal voltages were specified in cooperation with the ITER power supply group. They consist of several subsequent failures in the switching system and a fault of the coil insulation. The failure case 1 consists of four specifications:

- The currents of all CS and PF coils were set to rated values of 45 kA, similar to the current behaviour during fast discharge shown in Fig. 6.11.
- The DC voltages at the coil terminals were set to medium level, which means terminal voltage without booster converter and switching network unit. The voltages of the PF 3 coil were set to +3.75 kV at the positive terminal and -3.75 kV at the negative terminal. The terminal voltages of the PF 6 coil were set to ±0.75 kV.
- A fast discharge of the CS PF coil system was started under defined conditions and failures occurred in the free wheeling and power supply switches of particular coils. Consequently, the converter voltage and current were still on the coil terminals during fast discharge.
- An earth fault occurred at the negative coil terminal due to the cumulative voltages of the AC/DC converter and terminal voltages during fast discharge.

The voltage waveforms at the positive and negative terminals of the PF 3 coil in failure case 1 are shown in Fig. 6.16. The DC voltages at the coil terminals supplied by the main and vertical stabilisation converter before the ignition of the thyristor in the fast discharge unit (FDU) at 27 ms were defined to be ±3.75 kV in failure case 1. Compared to this value, the DC voltages before fast discharge without earth fault of about ±22 V were negligible. The earth fault at the negative terminal occurred at the maximum voltage of the first half-wave. The voltage dropped to 0 V after oscillations, which were caused by the 1 Ω earth fault resistance, and remained at this value until the end of fast discharge of the coil.

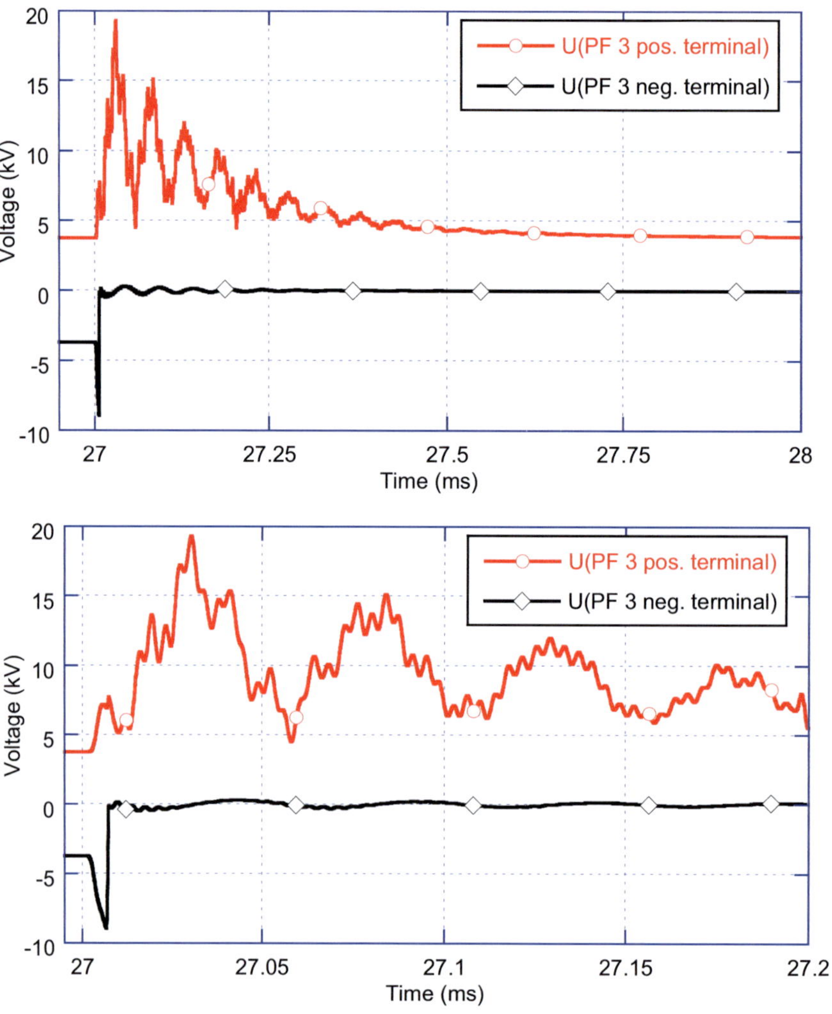

Fig. 6.16: Voltage waveforms at the positive and negative terminals of the PF 3 coil in failure case 1 in the upper part on a time scale between 26.9 ms and 28.0 ms and in the lower part on a time scale between 26.99 ms and 27.20 ms. Maximum time of sampling steps is 0.2 μs.

The oscillations at the positive terminal consisted of three interfering voltage waveforms. The first voltage waveform was caused by the ignition of the thyristor at 27 ms of fast discharge. The second was caused by an earth fault at the negative terminal 6.7 µs after the ignition of thyristor and the resultant displacement of DC voltage to the positive terminal of the PF 3 coil. The third voltage waveform was caused by the displacement of the voltage waveform between the coil terminals during fast discharge to the positive terminal due to an earth fault at the negative terminal 6.7 µs after the ignition of the thyristor.

The maximum voltage at the positive terminal of the PF 3 coil during failure case 1 was calculated to be 19.3 kV with a rise time of 15.5 µs. The interfering voltage oscillations decreased after about 0.5 ms to a negligible value and the voltage at the positive terminal consisted of DC voltage waveforms caused by converter voltages and the DC voltage of fast discharge after the commutation of coil current to the discharge resistor. The voltage at the positive terminal during fast discharge without fault has a negative value after the commutation of coil current to the discharge resistor due to Faraday's law of induction. On the other hand, the DC voltages supplied by the AC/DC converters of the coils are positive at the positive terminal. Due to the relatively high amplitude of the voltage at the positive terminal of 7.5 kV supplied by the AC/DC converters after the earth fault and the relatively low negative voltage of -3.7 kV at the positive terminal caused by the discharge of coil current, voltage at the positive terminal of the PF 3 coil in failure case 1 after the commutation of coil current to the discharge resistor remained positive at about 3.8 kV under steady-state conditions after 28 ms.

The maximum voltage at the negative terminal of the PF 3 coil during failure case 1 was calculated to be -9.0kV with a rise time of 3.5 µs. The voltage amplitude consisted of -5.3 kV caused by the ignition of the thyristor for discharge of the counterpulse capacitor and of the DC voltage of -3.75 kV caused by the AC/DC converter of the coil. An earth fault occurred at the negative terminal at a maximum voltage of the first half-wave. The fall time of the voltage during the earth fault was calculated to be 0.3 µs using the 10 % to 90 % principle.

The behaviour of the voltages at the positive and negative terminals of the PF 3 coil during failure case 1 was analysed by Fast Fourier Transformation (FFT) and is shown in Fig. 6.17. The behaviour of the voltage at the positive terminal after FFT shows a high proportion of voltage waveforms in the low-frequency range due to the DC voltage of the AC/DC converters. The second peak was calculated at 19.0 kHz, which is close to the resonance frequency of the PF 3 coil with unsymmetrical grounding at 20.7 kHz. Unsymmetrical grounding is caused by an earth fault, if the negative terminal has direct connection to the ground. The third and fourth peak values were calculated to be of 90 kHz and 262 kHz, respectively, which correspond to the peak value calculated during fast discharge without earth fault at 92 kHz and 263 kHz, respectively.

The behaviour of the voltage at the negative terminal after FFT shows high values of voltage in frequency range up to 50 kHz mainly due to the DC voltage of the AC/DC converters. The voltage decreases with increasing frequency, but still reaches 10 % of the maximum voltage at a frequency of about 220 kHz. The voltage behaviour at the negative terminal of the PF 3 coil in failure case 1 can be approximated by a unit step function superposed by a voltage impulse and both behaviours are comparable in the time domain as well as in the frequency domain.

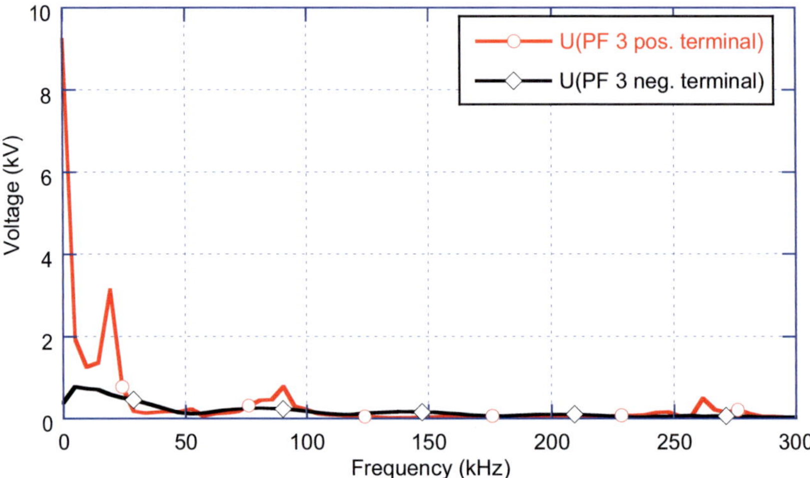

Fig. 6.17: FFT of voltage waveforms at the positive and negative terminals of the PF 3 coil in failure case 1 on the time scale between 26.99 ms and 27.20 ms.

The failure case 1 was calculated for the PF 6 coil with the same definitions as for PF 3 coil. However, the voltage at the coil terminals of PF 6 coil was set to 1.5 kV between the terminals, again supplied by the main AC/DC converter. The DC converter voltage at the terminals of the PF 6 coil was lower than at the terminals of the PF 3 coil, because the power supply circuit of the PF 6 coil consisted of one main converter without any vertical stabilisation converter to supply additional voltage to the terminals of the PF 3 coil. The voltage waveforms at the positive and negative terminals of the PF 6 coil in failure case 1 are shown in Fig. 6.18. The DC voltage at the coil terminals before ignition of the thyristor in the fast discharge unit was defined to be 1.5 kV between the terminals or ±0.75 kV at the positive and negative terminals, respectively. The DC voltage at the terminals of the PF 6 coil before fast discharge without earth fault was calculated to be ±23 V, as shown in Fig. 6.14. The earth fault at the negative terminal occurred at the maximum voltage of the first half-wave. The DC voltage of the AC/DC converter of 1.5 kV was displaced to the positive terminal after the earth fault.

Maximum voltage at the positive terminal was calculated to be 8.5 kV with a rise time of 12.3 µs. The value of 2.3 kV was defined to be the amplitude of the voltage for the calculation of the rise time with the 10 % to 90 % principle. The high harmonic voltage oscillations decreased to negligible values 0.8 ms after the ignition of the thyristor in the fast discharge unit at 27 ms. The voltage at the positive terminal of the PF 6 coil of -2.7 kV under steady-state conditions after the commutation of coil current to the fast discharge resistor after about 28 ms became negative, because the negative voltage of -4.2 kV caused by the discharge resistor was higher than the positive voltage of 1.5 kV caused by the main converter of the PF 6 coil.

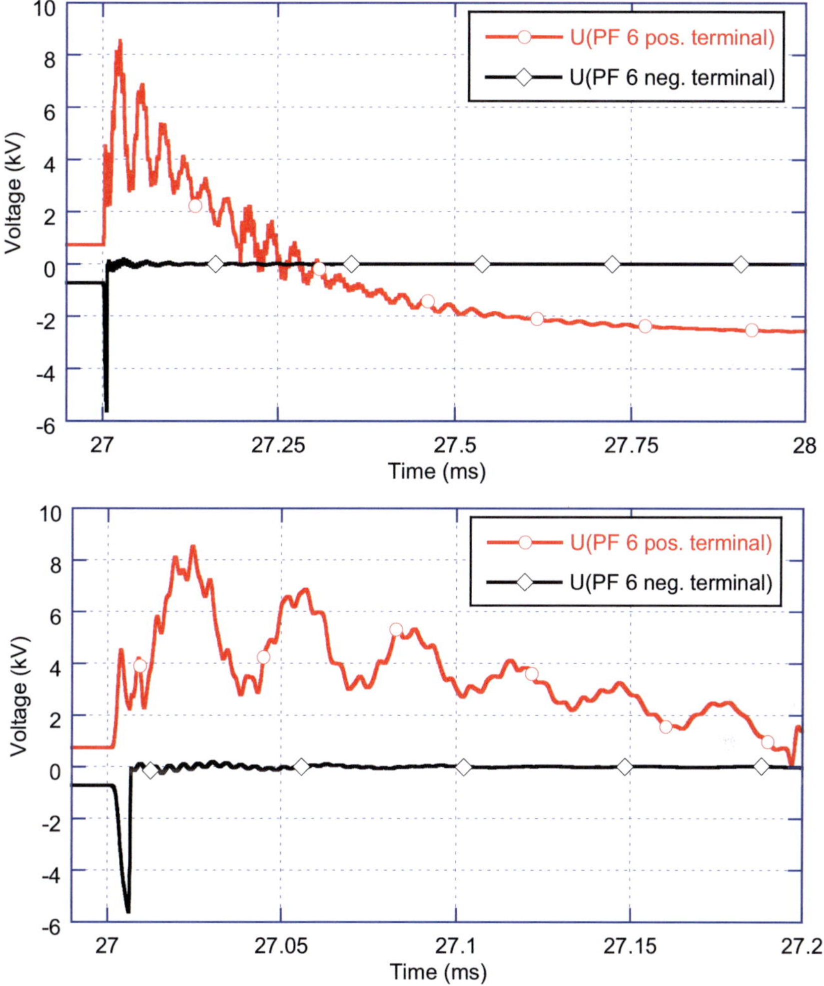

Fig. 6.18: Voltage waveforms at the positive and negative terminals of the PF 6 coil during failure case 1 in the upper part on the time scale between 26.9 ms and 28.0 ms and in the lower part on the time scale between 26.99 ms and 27.20 ms. Maximum time of sampling steps is 0.2 µs.

The maximum voltage at the negative terminal of the PF 6 coil in failure case 1 was calculated with the network model of the CS PF coil system to be -5.6 kV with a rise time of 2.9 µs. The maximum voltage consisted of the voltage of -4.9 kV caused by the discharge of the counterpulse capacitor starting at 27 ms and the DC voltage of -0.75 kV supplied by the main converter of the PF 6 coil power supply. Earth fault occurred at the negative terminal at maximum voltage of the first half-wave. The fall time of voltage during the earth fault was calculated to be 0.3 µs.

The behaviour of the voltages at the positive and negative terminals of the PF 6 coil during failure case 1 was analysed by Fast Fourier Transformation (FFT) and is shown in Fig. 6.19. The high voltage in the frequency range lower than 10 kHz was caused by DC voltage supplied by the AC/DC converter. The second peak value was calculated to be 33 kHz and was close to the resonance frequency of about 29 kHz of the PF 6 coil with unsymmetrical grounding. The third peak was calculated to be 190 kHz and comparable with the peak value of 189 kHz calculated during fast discharge without earth fault.

The behaviour of voltage at the negative terminal after FFT shows a high voltage in the frequency range up to 30 kHz mainly due to the DC voltage of the AC/DC converter before the earth fault. The voltage decreases slowly with increasing frequency, but still reaches 10 % of the maximum voltage at a frequency of about 236 kHz. The voltages of the PF 3 and PF 6 coils at the negative terminals in failure case 1 after FFT have a similar behaviour. The voltage waveforms at the negative terminal of the PF 3 coil have higher voltage amplitudes due to the additional voltage of the vertical stabilisation converter that is connected in series with the main converter of the power supply circuit of the PF 3 coil.

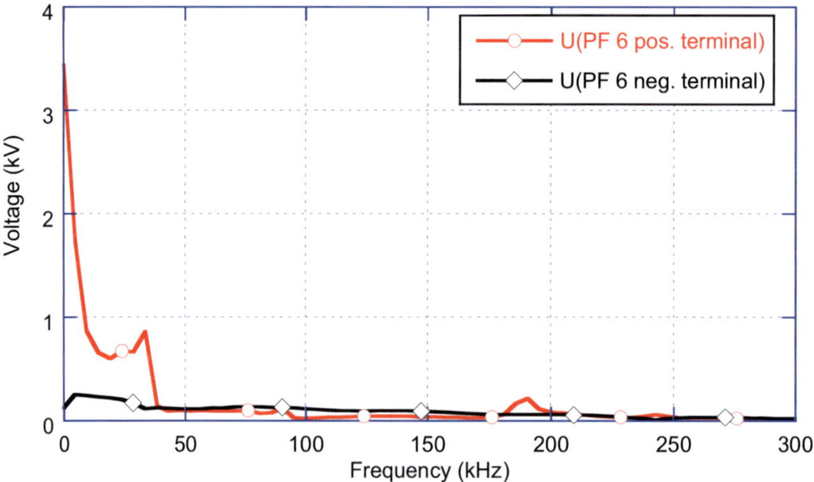

Fig. 6.19: FFT of voltage waveforms at the positive and negative terminals of the PF 6 coil in failure case 1 on the time scale between 26.99 ms and 27.20 ms.

6.2.4   Failure Case 2

Failure case 2 was also specified in cooperation with the ITER power supply group and consists of several subsequent failures in the switching system and a fault of coil insulation. The main difference to failure case 1 lies in the current and voltage levels at the beginning of failure case 2. The currents were set lower and the voltages higher. The remaining specifications and the order of the faults were the same as in failure case 1. The failure case 2 consists of four specifications:

- The currents of all CS and PF coils were set to 10 kA.
- The DC voltages at the coil terminals were set to the maximum level, including the voltage of the booster converter and switching network unit. The voltages of the PF 3 coil were set to +5.85 kV at the positive terminal and -5.85 kV at the negative terminal. The terminal voltages of the PF 6 coil were set to ±5.75 kV.
- A fast discharge of the CS PF coil system was started under defined conditions and failures occurred in the free wheeling and power supply switches of a particular coil, thus, the converter voltage and current were still driven through the coil during fast discharge.
- An earth fault occurred at the negative coil terminal due to the cumulative voltages of the AC/DC converter and terminal voltages during fast discharge.

The coil current was reduced to 10 kA and led to a shorter current discharge time constant. On the other hand, the level of the DC voltage at the coil terminals was increased, resulting in higher voltages at the terminals of both coils.

The behaviour of voltage waveforms at the positive and negative terminals of the PF 3 coil in failure case 2 are shown in Fig. 6.20. The DC voltage at the coil terminals in failure case 2 supplied by the main, booster and vertical stabilisation converters before the ignition of the thyristor in the fast discharge unit at 27.0 ms was defined to be ±5.85 kV. The earth fault at the negative terminal occurred at the maximum voltage of the first half-wave.

The maximum voltage at the positive terminal of the PF 3 coil was calculated to be 25.4 kV with a rise time of 15.8 µs. The high harmonic oscillations decreased at about 27.5 ms to negligible values. Consequently, voltage reached 10.9 kV under steady-state conditions. The terminal voltage was calculated from the positive DC voltage of 11.7 kV supplied by the AC/DC converters and the negative voltage caused by the discharge resistance of -0.8 kV. The voltage caused by the discharge resistor in failure case 2 was lower than during fast discharge without an earth fault due to the lower coil currents specified for this calculation scenario.

The maximum voltage at the negative terminal was calculated to be -11.1 kV with a rise time of 3.5 µs. The maximum voltages calculated at both terminals of the PF 3 coil in failure case 2 were also the maximum voltages calculated for all four calculation scenarios.

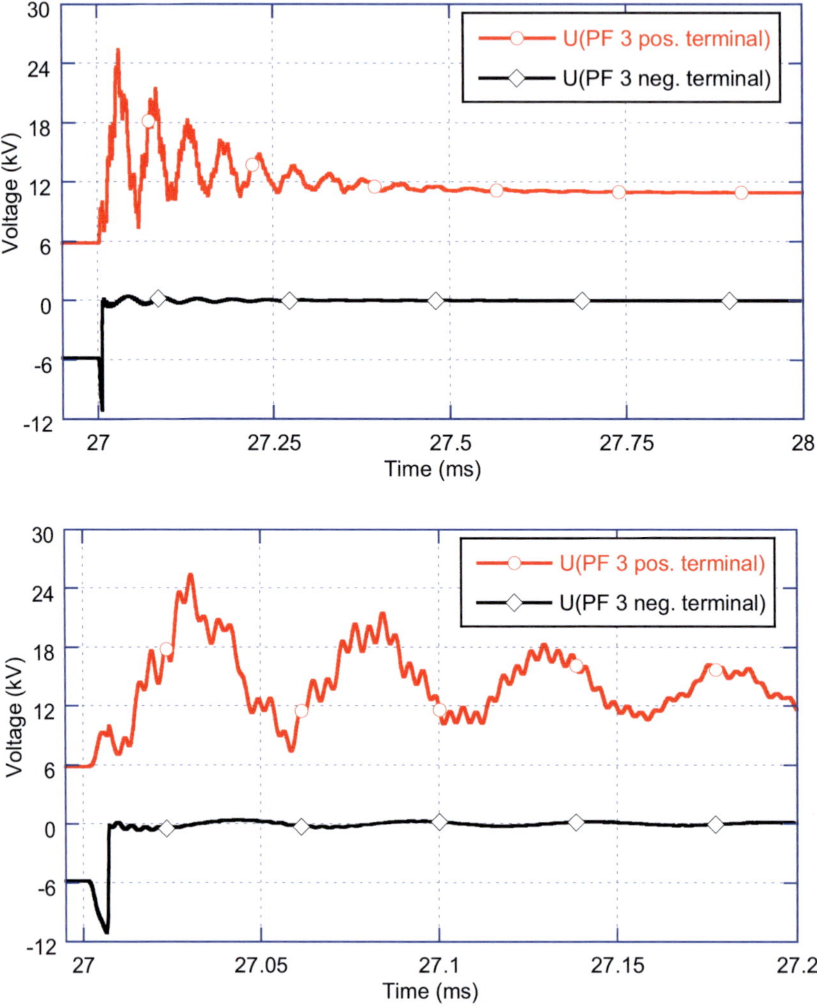

Fig. 6.20: Voltage waveforms at the positive and negative terminals of the PF 3 coil during failure case 2
in the upper part on a time scale between 26.9 ms and 28.0 ms and in the lower part on a time
scale between 26.99 ms and 27.20 ms. Maximum time of sampling steps 0.2 μs.

The behaviour of voltages at the positive and negative terminals of the PF 3 coil during failure case 2 was analysed by Fast Fourier Transformation (FFT) and is shown in Fig. 6.21. The high voltage in the frequency range lower than 4 kHz was caused by the high DC voltage supplied by the AC/DC converters. The second peak value was calculated to be at 19.0 kHz, which is close to the resonance frequency of the asymmetrically grounded PF 3 coil of 20.7 kHz. The third and fourth peak values were calculated to be 91 kHz and 262 kHz, respectively. These calculated frequencies are close to the frequencies calculated after the FFT of voltages during fast discharge of 92 kHz and 263 kHz.

The behaviour of voltage at the negative terminal after FFT shows high values in the frequency range up to 50 kHz mainly caused by the DC voltage of the AC/DC converters. The voltage decreases with increasing frequency faster than the voltage in failure case 1 and reaches 10 % of the maximum voltage at a frequency of about 110 kHz.

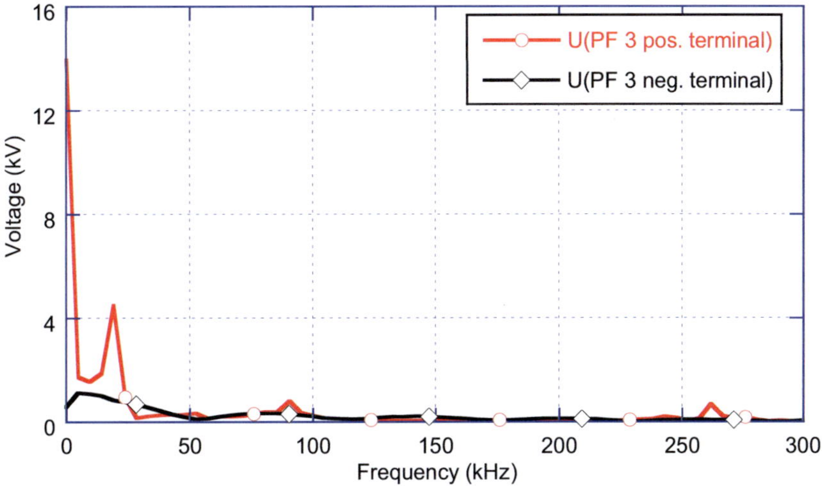

Fig. 6.21: FFT of voltage waveforms at the positive and negative terminals of the PF 3 coil in failure case 2 on the time scale between 26.99 ms and 27.20 ms.

The failure case 2 was calculated for the PF 6 coil with the same specifications as for the PF 3 coil. However, the voltage at the coil terminals of the PF 6 coil was set to 11.5 kV between the terminals. A maximum DC voltage of 1.5 kV was supplied by the main AC/DC converter and a maximum DC voltage of 10 kV by the switching network unit (SNU). The voltage waveforms at the positive and negative terminals of the PF 6 coil in failure case 2 are shown in Fig. 6.22. The DC voltage at the coil terminals before ignition of the thyristor in the fast discharge unit was defined to be 11.5 kV between the terminals or ±5.75 kV at the positive and negative terminals to ground. The earth fault at the negative terminal occurred at the maximum voltage of the first half-wave. The DC voltage of 11.5 kV between the coil terminals was displaced to the positive terminal after the earth fault and caused additional voltage oscillations.

Maximum voltage at the positive terminal was calculated to be 23.4 kV with a rise time of 10.4 µs. The value of 7.2 kV was defined to be the amplitude of the voltage for calculation of the rise time with the 10 % to 90 % principle. The high harmonic voltage oscillations decreased to negligible values 0.8 ms after the ignition of the thyristor in the fast discharge unit at 27 ms. The voltage at the positive terminal of the PF 6 coil was calculated to be 10.6 kV under steady-state conditions after the commutation of coil current to the fast discharge resistor at about 28 ms. The voltage remained positive due to the high positive DC voltage supplied by the main converter and switching network unit.

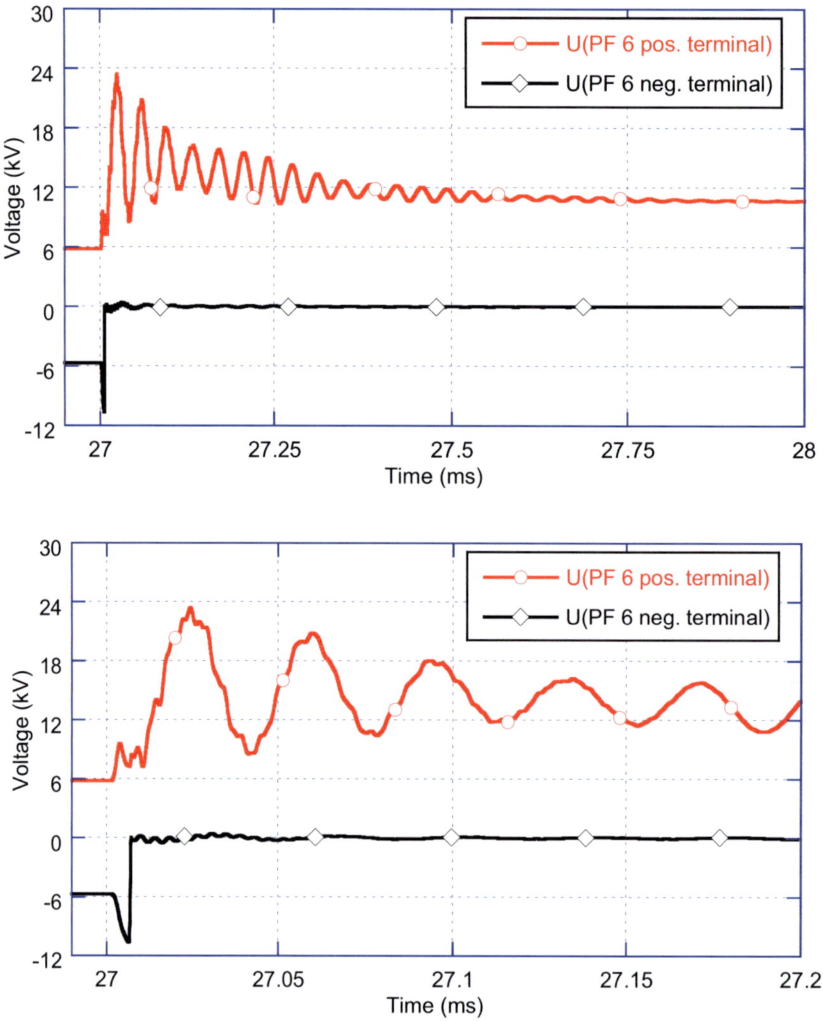

Fig. 6.22: Voltage waveforms at the positive and negative terminals of the PF 6 coil during failure case 2 in the upper part on a time scale between 26.9 ms and 28.0 ms and in the lower part on a time scale between 26.99 ms and 27.20 ms. Maximum time of sampling steps 0.2 μs.

The maximum voltage at the negative terminal of the PF 6 coil in failure case 2 was calculated with the network model of the CS PF coil system to be -10.7 kV with a rise time of 3.0 µs. The maximum voltage consisted of the voltage of -4.9 kV caused by the discharge of the counterpulse capacitor starting at 27 ms and the DC voltage of -5.75 kV supplied by the main converter and switching network unit of the PF 6 coil power supply. The earth fault occurred at the negative terminal at maximum voltage of the first half-wave. The fall time of voltage during the earth fault was calculated to be 0.3 µs.

The behaviour of voltages at the positive and negative terminals of the PF 6 coil during failure case 2 was analysed by Fast Fourier Transformation (FFT) and is shown in Fig. 6.23. The high voltage in the frequency range lower than 5 kHz was caused by the DC voltage supplied by the AC/DC converter and switching network unit. The second peak value at 29 kHz was comparable to the resonance frequency of about 29 kHz of PF 6 coil with unsymmetrical grounding. The third peak was calculated to be 191 kHz and reached 2 % of the maximum amplitude of the voltage calculated at DC only.

The behaviour of voltage at the negative terminal after FFT shows high values in the frequency range up to 40 kHz mainly caused by the DC voltage of the AC/DC converter and switching network unit before the earth fault. The voltage decreases slowly with increasing frequency, but still reaches 10 % of the maximum voltage at a frequency of about 200 kHz. The voltages of the PF 3 and PF 6 coils at the negative terminals in failure case 2 after FFT have similar behaviours due to the high DC voltages at the negative terminals of both coils.

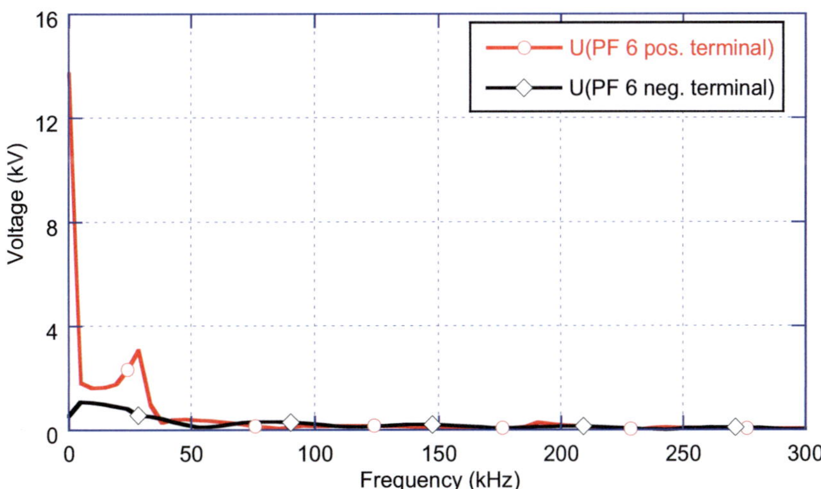

Fig. 6.23: FFT of voltage waveforms at the positive and negative terminals of the PF 6 coil in failure case 2 on a time scale between 26.99 ms and 27.20 ms.

6.3   Internal Voltage Distribution of the Coils

The internal voltage distribution was calculated to analyse the amplitudes and waveforms of voltages on the ground, layer and turn insulations within the coils. The calculated maximum voltages will be used for definition of amplitudes and waveforms for high-voltage tests of the coil insulation during the manufacturing process. The voltage waveforms calculated at the terminals of the PF 3 and PF 6 coils with the network model of the CS PF coil system were used as excitations at the coil terminals of the detailed network models of the PF 3 and PF 6 coils. The internal voltage distribution was calculated for four calculation scenarios consisting of voltage oscillations with different frequencies. Three network models for low, resonance and high frequencies were chosen from the generated detailed network models used for the calculation of the resonance frequency of the PF 3 and PF 6 coils. The calculations were made for discrete frequencies due to frequency-dependent inductances and because voltage excitations with different frequencies were calculated at the terminals of the coils. The detailed network model of the PF 3 coil for low frequencies was built with inductances calculated at 0.51 kHz. It represented slow excitations mainly occurring in the reference scenario. The resonance frequency network model of the PF 3 coil without instrumentation cables was built with inductances calculated at 25.5 kHz and represented excitations in the range of the resonance frequency of the coil. The excitations and oscillations in failure cases 1 and 2 consisted of voltages close to resonance frequency. The detailed network model of PF 3 coil for high frequencies was built at 200 kHz and represented fast excitations occurring during fast discharge and also in both failure cases. The detailed network models for 0.55 kHz, 32 kHz and 200 kHz were used for calculation of internal voltage within the PF 6 coil. The calculations of internal voltage distribution were made for all four calculation scenarios with the three detailed network models of each coil to analyse the influence of the frequency dependency of the inductances on the maximum voltage on different kinds of high-voltage insulation. The voltage waveforms and maximum voltages within the PF 3 and PF 6 coils calculated for the four calculation scenarios are described in the following sections.

6.3.1   Reference Scenario

The reference scenario represents the rated operation of the coils and was specified to have a total length of 1800 s. The voltage waveforms at the coil terminals were supplied by the AC/DC converter to ensure the coil currents required for operation of ITER coils. The voltages at the coil terminals were calculated with the defined current behaviours of CS and PF coils as specified in [Mit09]. The maximum voltages at the coil terminals were calculated during the first 1.4 s of the reference scenario because of the current alternations necessary for the ignition of the plasma. Thus, internal voltage distribution was analysed in the first 1.4 s of the reference scenario with the detailed network models of the PF 3 and PF 6 coils at low, resonance and high frequency.

The voltage behaviour calculated at the terminals of the PF 3 and PF 6 coils during the reference scenario mainly consisted of DC voltage at different levels with transition times between the voltage levels of 2.2 ms and 4.0 ms, respectively. The rise time of 2.2 ms corresponded to the rise time of sine voltage at about 55 Hz. The voltage calculated within the PF 3 coil followed voltage waveforms at the coil terminals due to the relatively slow rise times of excitations. This behaviour led to a linear voltage distribution within the coils during the reference scenario. The maximum voltage on ground insulation was calculated on the first turn at the positive terminal of the PF 3 coil between turn P01_C12 and ground, shown in Fig. 6.24.

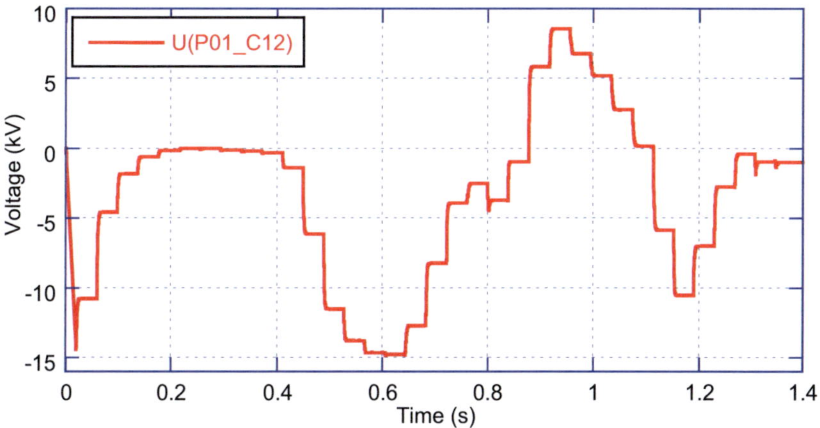

Fig. 6.24: Voltage waveforms of the calculated maximum voltage of ground insulation between conductor P01_C12 of the PF 3 coil and ground during the reference scenario. Maximum time of sampling steps 1 ms.

Due to the relatively slow rise time of the excitations and the resultant linear voltage distribution during the reference scenario within the PF 3 and PF 6 coils, the calculated internal voltages on layer and turn insulation were relatively low. The voltages on ground, layer and turn insulations of the PF 3 coil are summarised in Tab. 6.5. The time when the calculated maximum voltage occurred for all three insulation types corresponded to the time when the maximum voltage was calculated at the terminals of the PF 3 coil at 608 ms. The rise times of the maximum voltages for different kinds of insulation were also calculated to be 2.2 ms. The rise time was calculated with the 10 % to 90 % principle. Tab. 6.5 also includes the value for the start rise voltage used for the calculation of the rise time.

The calculated internal voltage behaviour of the PF 6 coil in the reference scenario followed the relatively slow excitations at the coil terminals. The maximum voltages calculated on ground, layer and turn insulations of the PF 6 coil are summarised in Tab. 6.6. The maximum voltages were calculated at 20 ms when the current of the PF 6 coil had its highest derivative and, hence, the highest terminal voltage for preparation of plasma ignition.

Tab. 6.5:   Maximum voltages calculated with the network models of the PF 3 coil for 0.51 kHz, 25.5 kHz and 200 kHz during the reference scenario. Due to relatively slow excitations and linear voltage distribution within the coils, the results for the 0.51 kHz, 25.5 kHz and 200 kHz network models were the same.

	Terminal to ground	Insulation turn to ground	Insulation between layers	Insulation between turns
Value of maximum voltage	-14.9 kV	-14.9 kV	-2.6 kV	0.9 kV
Location of maximum voltage	Positive terminal	P01_C12 to ground	P12_C01 to P13_C01	P01_C01 to P01_C02
Time of maximum voltage	607 ms	607 ms	608 ms	608 ms
Rise time of maximum voltage	2.2 ms	2.2 ms	2.2 ms	2.2 ms
Start rise voltage	-13.8 kV	-13.8 kV	-2.4 kV	0.8 kV

Tab. 6.6:   Maximum voltages calculated with the network models of the PF 6 coil for 0.55 kHz, 32 kHz and 200 kHz during the reference scenario. Due to relatively slow excitations and linear voltage distribution within the coils, the results for the 0.55 kHz, 32 kHz and 200 kHz network models were the same.

	Terminal to ground	Insulation turn to ground	Insulation between layers	Insulation between conductors
Value of maximum voltage	7.2 kV	7.2 kV	-2.6 kV	-0.9 kV
Location of maximum voltage	Negative terminal	P16_C27 to ground	P14_C02 to P15_C02	P04_C26 to P04_C27
Time of maximum voltage	20 ms	20 ms	20 ms	20 ms
Rise time of maximum voltage	4ms	4 ms	4 ms	4 ms
Start rise voltage	6.3 kV	6.3 kV	-2.3 kV	-0.8 kV

6.3.2 Fast Discharge

Fast discharge will be started, if a failure occurs in one of the superconducting coils or in their electrical, cryogenic and vacuum supply systems. All CS and PF coils will be discharged simultaneously due to their magnetic and partly galvanic coupling [DDD06m]. The current discharge time constant for PF coils was specified to be 14 s [DDD06c]. Due to the different inductances of the PF coils, the fast discharge resistors had different values to reach the defined current discharge time constant of 14 s. The DC currents of the coils had to be commutated from the free wheeling circuit of the fast discharge unit (FDU) to the discharge resistor by the discharge of the preloaded counterpulse capacitor. The commutation procedure caused fast voltage excitations at the coil terminals, described in detail in section 6.2.2. The high amplitude of the excitation voltages on the coil terminals and their fast rise time caused oscillations within the coils, as shown in Fig. 6.25 using the voltage waveform on ground insulation of PF 3 coil as an example.

The discharge of the counterpulse capacitor was started by the ignition of thyristor at 27 ms. The maximum voltage during fast discharge at the positive terminal of the PF 3 coil was calculated at 27.028 ms and at the negative terminal at 27.007 ms. The maximum voltage on the ground insulation shown in Fig. 6.25 was calculated to be reached at 27.301 ms on turn P13_C12 located in the lower part of the PF 3 coil consisting of 16 pancakes (P16) in total. The calculated voltage waveform on the ground insulation shows an increase of the oscillations to the maximum calculated voltage of 12.3 kV. The maximum voltages on ground, layer and turn insulations calculated during fast discharge within the PF 3 coil are summarised in Tab. 6.7.

The maximum voltages within the PF 3 coil were calculated with network models for low, resonance and high frequencies at 0.51 kHz, 25.5 kHz and 200 kHz, respectively. The maximum voltages on the ground, layer and turn insulation calculated for all three network models of the PF 3 coil during fast discharge are shown in red in Tab. 6.7. The highest voltage on the ground insulation was calculated to be 12.3 kV with the 200 kHz model for fast excitations. The absolute maximum voltage calculated to be -12.1 kV on the ground insulation with the 0.51 kHz network model is only 0.2 kV lower. The maximum voltage calculated on the ground insulation with 25.5 kHz model was nearly two times lower. For the layer and turn insulations the maximum voltages were calculated with the low-frequency model. The values calculated with the 200 kHz model were in the same range with a maximum difference of 17 %. The maximum voltages calculated with the resonance frequency model were up to two times lower than the maximum voltage calculated with the low-frequency model. The excitation of the PF 3 coil during fast discharge mainly consisted of excitations at low and high frequencies. Hence, the maximum voltage within the coil was also calculated with detailed network models for low and high frequencies. The waveforms of the maximum voltages calculated on the layer and turn insulations of the PF 3 coil are shown in Annex A.4. The rise time of the maximum voltages within the coil was calculated to be in the range between 3.0 µs and 3.8 µs and corresponded to the rise time of 3.5 µs of the maximum voltage at the coil terminals. This rise time is comparable to the rise time of sine voltage at frequency of 95 kHz calculated with 10 % to 90 % of the peak-to-peak voltage.

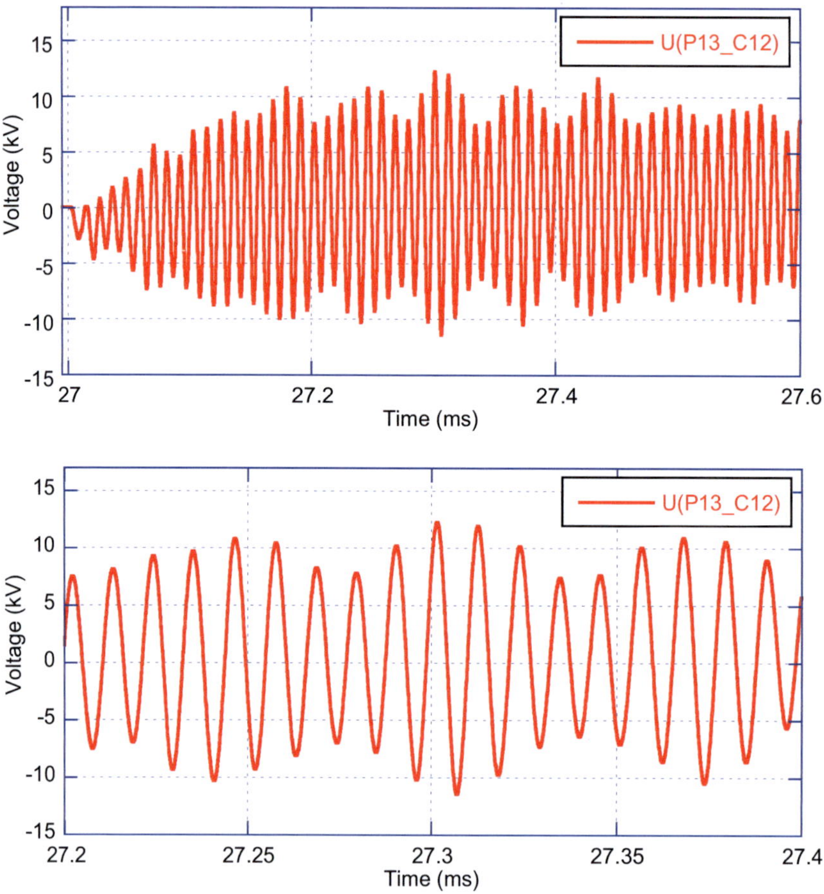

Fig. 6.25: Voltage waveforms of maximum voltage calculated on the ground insulation between turn P13_C12 and ground during fast discharge. Calculated with the 200 kHz detailed network model of the PF 3 coil. Maximum time of sampling steps 0.1 µs. Upper part on the time scale between 26.99 ms and 27.60 ms, lower part on the time scale between 27.2 ms and 27.4 ms.

The maximum voltages calculated within the PF 6 coil during fast discharge were lower than the voltage within the PF 3 coil, although the excitation voltages of both coils are comparable in amplitude and rise time, -5.3 kV and 3.5 µs for the PF 3 coil and -5.0 kV and 2.7 µs for the PF 6 coil. The maximum voltages within the PF 6 coil are listed in Tab. 6.8 and were calculated with detailed network models for low 0.55 kHz, resonance 32.0 kHz and high frequency 200 kHz. The maximum voltages calculated on the ground, layer and turn insulations of the PF 6 are shown in red. The rise times of the internal maximum voltages of the PF 6 coil are in the range between 0.8 µs and 2.4 µs. These rise times corresponded to the excitation voltage at the positive and negative terminals of the PF 6 coil during fast discharge. The maximum voltages within the PF 3 and PF 6 coils were calculated with the low- and high-frequency models.

Tab. 6.7: Maximum voltages calculated with the detailed network models of the <u>PF 3</u> coil during fast discharge. Maximum voltages calculated on particular insulations are shown in red.

		Terminal to ground	Insulation turn to ground	Insulation between layers	Insulation between turns
0.51 kHz network model	Value of maximum voltage	-5.3 kV	-12.1 kV	-13.0 kV	10.0 kV
	Location of maximum voltage	Negative terminal	P01_C06 to ground	P01_C08 to P02_C08	P01_C08 to P01_C09
	Time of maximum voltage	27.007 ms	27.523 ms	27.154 ms	27.138 ms
	Rise time of maximum voltage	3.5 µs	2.97 µs	3.05 µs	3.4 µs
	Start rise voltage	0.0 kV	8.7 kV	12.9 kV	-9.0 kV
25.5 kHz network model	Value of maximum voltage	-5.3 kV	7.0 kV	8.2 kV	-4.9 kV
	Location of maximum voltage	Negative terminal	P01_C02 to ground	P08_C02 to P09_C02	P14_C10 to P14_C11
	Time of maximum voltage	27.007 ms	27.105 ms	27.068 ms	27.104 ms
	Rise time of maximum voltage	3.5 µs	3.5 µs	3.8 µs	3.4 µs
	Start rise voltage	0.0 kV	-3.7 kV	-7.1 kV	3.6 kV
200 kHz network model	Value of maximum voltage	-5.3 kV	12.3 kV	-11.3 kV	-8.3 kV
	Location of maximum voltage	Negative terminal	P13_C12	P08_C02 to P09_C02	P04_C11 to P04_C12
	Time of maximum voltage	27.007 ms	27.301 ms	27.290 ms	27.251 ms
	Rise time of maximum voltage	3.5 µs	3.3 µs	3.3 µs	3.5 µs
	Start rise voltage	0.0 kV	-10.2 kV	11.2 kV	7.0 kV

Tab. 6.8: Maximum voltages calculated with the detailed network models of the PF 6 coil during fast discharge. Maximum voltages calculated on particular insulations are shown in red.

		Terminal to ground	Insulation turn to ground	Insulation between layers	Insulation between turns
0.55 kHz network model	Value of maximum voltage	-5.0 kV	-8.4 kV	7.6 kV	4.9 kV
	Location of maximum voltage	Negative terminal	P16_C19 to ground	P14_C19 to P15_C19	P01_C13 to P01_C14
	Time of maximum voltage	27.027 ms	27.017 ms	27.279 ms	27.174 ms
	Rise time of maximum voltage	2.7 µs	2.0 µs	1.7 µs	1.7 µs
	Start rise voltage	1.0 kV	3.0 kV	-6.3 kV	-1.5 kV
32 kHz network model	Value of maximum voltage	-5.0 kV	7.2 kV	5.8 kV	4.9 kV
	Location of maximum voltage	Negative terminal	P01_C17 to ground	P01_C17 to P02_C17	P16_C22 to P16_C23
	Time of maximum voltage	27.027 ms	27.015 ms	27.020 ms	27.036 ms
	Rise time of maximum voltage	2.7 µs	1.0 µs	0.8 µs	1.3 µs
	Start rise voltage	1.0 kV	-3.4 kV	-5.4 kV	-2.9 kV
200 kHz network model	Value of maximum voltage	-5.0 kV	-9.8 kV	-7.4 kV	7.5 kV
	Location of maximum voltage	Negative terminal	P03_C01 to ground	P02_C09 to P03_C09	P16_C24 to P16_C25
	Time of maximum voltage	27.027 ms	27.348 ms	27.353 ms	27.077 ms
	Rise time of maximum voltage	2.7 µs	1.7 µs	1.8 µs	2.4 µs
	Start rise voltage	1.0 kV	5.6 kV	2.7 kV	-5.0 kV

6.3.3   Failure Case 1

The two failure cases represent possible scenarios of subsequent failures in the coil power supply system and were defined in cooperation with the power supply group of ITER. Failure case 1 was defined to be an earth fault during fast discharge at the negative terminal of the PF 3 and PF 6 coil, respectively. The coils were driven with rated currents of 45 kA and with voltage between the coil terminals without booster converter (BC) and without switching network units (SNU), 7.5 kV for the PF 3 coil and 1.5 kV for the PF 6 coil. The coil terminals were grounded symmetrically by a resistor network. Hence, voltage at the terminals of the PF 3 coil was ±3.75 kV and at the terminals of the PF 6 coil ±0.75 kV. Due to the specified failure in free wheeling and power supply switches, the DC voltage and current generated by the AC/DC converters was applied to the coil terminals during failure case 1. An earth fault occurred at the maximum voltage of the first half-wave at the negative terminal of the affected coil due to the cumulative voltage consisting of the DC voltage supplied by the converters and additional voltage by the commutation of current during fast discharge. The voltage at the negative terminal dropped after the earth fault to 0 kV and the voltage from the negative terminal was transferred to the positive terminal with an additional oscillation caused by the earth fault. The detailed voltage behaviour calculated at the terminals of the PF 3 and PF 6 coils in failure case 1 are shown in section 6.2.3.

The calculated voltage waveforms at the terminals of both coils were used as excitations at the terminals of the detailed network models of the PF 3 and PF 6 coils to calculate the internal voltage distribution. The behaviour of maximum voltage calculated on the ground insulation of the PF 3 coil during failure case 1 is shown in Fig. 6.26. The discharge of the preloaded counterpulse capacitor for commutation of coil current to the discharge resistor at 27 ms caused the first increase of the voltage from 3.75 kV to 8.75 kV on the ground insulation. The earth fault at the negative terminal occurred at 27.007 ms and caused a transition of the voltage from the negative to the positive terminal with additional oscillations of internal voltage. These oscillations led to a maximum calculated voltage of 23.4 kV on the ground insulation on the turn P01_C06 in the middle of the upper pancake of the PF 3 coil. The maximum voltage decreased to values lower than 10 kV about 300 µs after the earth fault. The rise time of the maximum voltage was calculated to be 4.1 µs with the 10 % to 90 % principle. The maximum voltages calculated within the PF 3 coil with network models for low, resonance and high frequencies in failure case 1 are listed in Tab. 6.9. The maximum voltages calculated in failure case 1 on the ground, layer and turn insulations are shown in red.

The maximum voltages on the ground insulation within the PF 3 coil were higher than the calculated voltages at the terminals. The maximum voltages within the PF 3 coil were calculated with detailed network models for low and resonance frequencies and corresponded to the voltage excitations calculated at the terminals of the PF 3 coil at frequencies lower than 4.0 kHz and at 19.0 kHz which is close to resonance frequency of the coil at 25.5 kHz.

The maximum voltages calculated within the PF 6 coil were lower than voltages within the PF 3 coil due to the lower DC voltages supplied by the main converter of its power supply circuit. The maximum voltages on the ground, layer and turn insulations of the PF 6 coil calculated with the detailed network models for low, resonance and high frequency are listed in Tab. 6.10. The maximum voltages were calculated in pancake 01 (P01) connected to the positive terminal of the PF 6 coil with the maximum terminal voltage in failure case 1. The voltages calculated within the coil on the ground insulation were higher than at the coil terminals of the PF 6 coil due to the internal oscillation and non-linear voltage distribution.

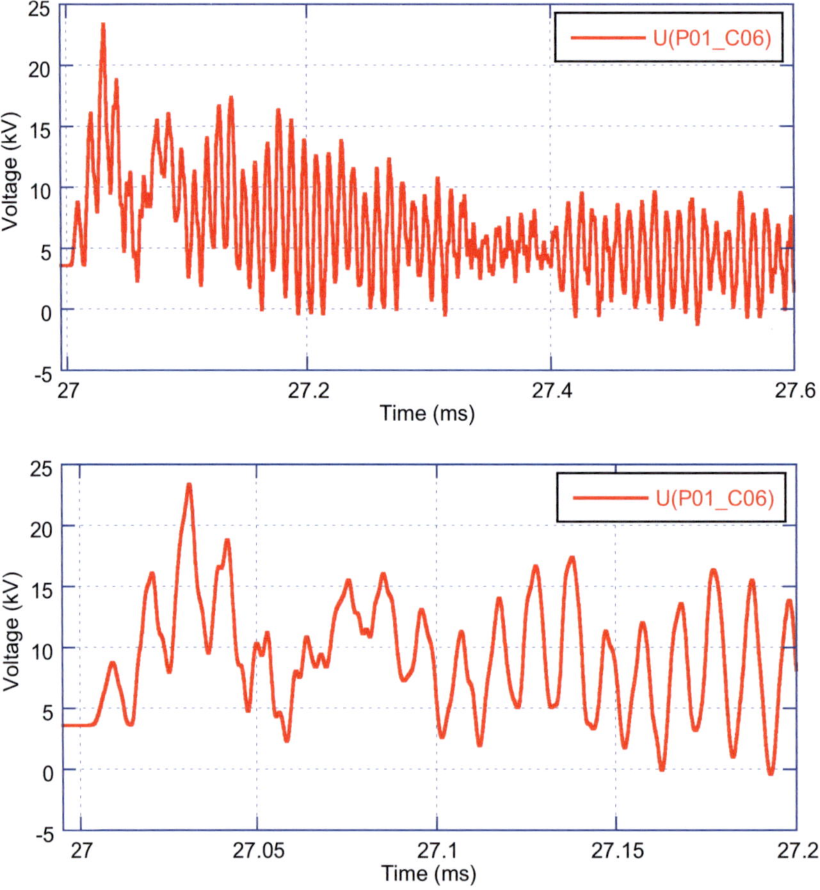

Fig. 6.26: Voltage waveforms of maximum voltage calculated on the ground insulation between conductor P01_C06 and ground in failure case 1 with the 0.51 kHz detailed network model of the PF 3 coil. Maximum time of sampling steps 0.1 µs. Upper part on the time scale between 26.99 ms and 27.60 ms, lower part on the time scale between 26.99 ms and 27.20 ms.

The maximum voltages within the PF 6 coil were calculated with detailed network models for resonance and high frequencies and are shown in red in Tab. 6.10. The voltages calculated with the other network models were in the same voltage range and showed the highest difference to the maximum voltages of 16 %.

Tab. 6.9: Maximum voltages calculated with the network models of the PF 3 coil in failure case 1. Maximum voltages calculated on particular insulations are shown in red.

		Terminal to ground	Insulation turn to ground	Insulation between layers	Insulation between turns
0.51 kHz network model	Value of maximum voltage	19.3 kV	23.4 kV	10.8 kV	7.6 kV
	Location of maximum voltage	Positive terminal	P01_C06 to ground	P02_C03 to P03_C03	P02_C09 to P02_C10
	Time of maximum voltage	27.031 ms	27.031 ms	27.080 ms	27.074 ms
	Rise time of maximum voltage	15.5 µs	4.1 µs	15.6 µs	14.9 µs
	Start rise voltage	5.1 kV	8.0 kV	-6.5 kV	-5.7 kV
25.5 kHz network model	Value of maximum voltage	19.3 kV	21.3 kV	14.0 kV	9.6 kV
	Location of maximum voltage	Positive terminal	P02_C01 to ground	P01_C06 to P02_C06	P02_C01 to P02_C02
	Time of maximum voltage	27.031 ms	27.031 ms	27.116 ms	27.310 ms
	Rise time of maximum voltage	15.5 µs	15.1 µs	1.1 µs	1.2 µs
	Start rise voltage	5.1 kV	5.0 kV	-7.2 kV	2.6 kV
200 kHz network model	Value of maximum voltage	19.3 kV	20.8 kV	8.4 kV	-8.4 kV
	Location of maximum voltage	Positive terminal	P01_C06 to ground	P15_C08 to P16_C08	P02_C04 to P02_C05
	Time of maximum voltage	27.031 ms	27.086 ms	27.029 ms	27.110 ms
	Rise time of maximum voltage	15.5 µs	5.1 µs	2.1 µs	2.3 µs
	Start rise voltage	5.1 kV	9.5 kV	-1.9 kV	2.0 kV

Tab. 6.10: Maximum voltages calculated with the network models of the <u>PF 6</u> coil in failure case 1. Maximum voltages calculated on particular insulations are shown in red.

		Terminal to ground	Insulation turn to ground	Insulation between layers	Insulation between turns
0.55 kHz network model	Value of maximum voltage	8.5 kV	10.1 kV	5.7 kV	4.5 kV
	Location of maximum voltage	Positive terminal	P01_C13 to ground	P15_C13 to P16_C13	P15_C05 to P15_C06
	Time of maximum voltage	27.025 ms	27.023 ms	27.485 ms	27.012 ms
	Rise time of maximum voltage	12.3 µs	10.4 µs	1.8 µs	2.5 µs
	Start rise voltage	2.3 kV	1.9 kV	-3.8 kV	-0.4 kV
32 kHz network model	Value of maximum voltage	8.5 kV	12.0 kV	5.9 kV	4.5 kV
	Location of maximum voltage	Positive terminal	P01_C13 to ground	P01_C15 to P02_C15	P14_C19 to P14_C20
	Time of maximum voltage	27.025 ms	27.024 ms	27.051 ms	27.462 ms
	Rise time of maximum voltage	12.3 µs	1.4 µs	1.3 µs	1.9 µs
	Start rise voltage	2.3 kV	3.7 kV	-2.2 kV	-1.5 kV
200 kHz network model	Value of maximum voltage	8.5 kV	11.1 kV	6.2 kV	4.9 kV
	Location of maximum voltage	Positive terminal	P01_C15 to ground	P01_C19 to P02_C19	P01_C17 to P01_C18
	Time of maximum voltage	27.025 ms	27.023 ms	27.337 ms	27.849 ms
	Rise time of maximum voltage	12.3 µs	1.3 µs	1.4 µs	1.2 µs
	Start rise voltage	2.3 kV	4.7 kV	-3.2 kV	-1.8 kV

6.3.4 Failure Case 2

Failure case 2 was specified to be an earth fault at the negative terminal of one coil during fast discharge similar to failure case 1. The difference between the two failure cases was the current and voltage levels at the coil terminals before the earth fault occurred. In failure case 2, the coil currents were set to 10 kA and the voltages at the terminals were set to the rated values of all converter and switching network units in the power supply circuits of the coils. Thus, the DC voltage at the terminals of the PF 3 coil was set to ±5.85 kV and at the terminals of the PF 6 coil to ±5.75 kV. Due to the specified failures in free wheeling and power supply switches, the DC voltage and current generated by the AC/DC converters were applied to the coil terminals during failure case 2. Fast discharge started for commutation of the coil current to the discharge resistor. The discharge of the preloaded counterpulse capacitor for current commutation started at 27 ms and caused a fast increase of the voltage at the terminals of the affected coil. The DC voltage supplied by the converters and switching network units of the coil circuits and the additional voltage during fast discharge increased the terminal voltage to maximum values in the four calculation scenarios. The earth fault occurred at the negative terminal of the affected coil at the maximum voltage of the first half-wave. The earth fault caused a voltage transition from the negative terminal to the positive terminal, followed by voltage oscillations within the coil.

The calculated voltage at the terminals of the PF 3 and PF 6 coils were included as voltage excitations in the detailed network models of both coils for calculation of internal voltage distribution. The voltage waveform calculated on the ground insulation of the PF 3 coil is shown in Fig. 6.27. The maximum voltages within the PF 3 coil calculated with low-, resonance and high-frequency network models in failure case 2 are summarised in Tab. 6.11. The maximum voltages calculated on the ground, layer and turn insulations of the PF 3 coil in failure case 2 are shown in red.

The maximum voltages within the PF 3 coil were calculated with the resonance frequency network model. The internal maximum voltage on the ground insulation was higher than the maximum voltage at the coil terminals due to the internal oscillations and non-linear voltage distribution. The rise time of the maximum voltage at the positive terminal with a 15.8 µs peak-to-peak value corresponds to the rise time of sine voltage at about 20 kHz and is close to the resonance frequency of the PF 3 coil at 25.5 kHz. The oscillations of the voltage waveform on the ground insulation at a frequency of about 20 kHz are shown in Fig. 6.27. Thus, the voltage excitations of the PF 3 coil in failure case 2 were close to the resonance frequency and maximum voltages within the PF 3 coil were calculated with the model for resonance frequency. The rise times of the maximum voltages varied from 1.5 µs to 16.5 µs depending on location and kind of insulation.

The maximum voltages calculated with low, resonance and frequency network models within the PF 6 coil in failure case 2 are listed in Tab. 6.12. The maximum voltage at the positive terminal of the PF 6 coil of 23.4 kV is 2 kV lower than at the terminals of the PF 3 coil. On the other hand, the maximum voltage of 30.1 kV on the ground insulation within the PF 6 coil was the highest voltage within both coils for the four calculation scenarios. The internal voltage calculated on the ground insulation of the PF 6 coil was higher than the maximum voltage at the coil terminals. The maximum voltages on layer and turn insulations within the PF 6 coil were lower than within the PF 3 coil. The waveforms of maximum voltages calculated on layer and turn insulations of the PF 3 coil are shown in Annex A.4.

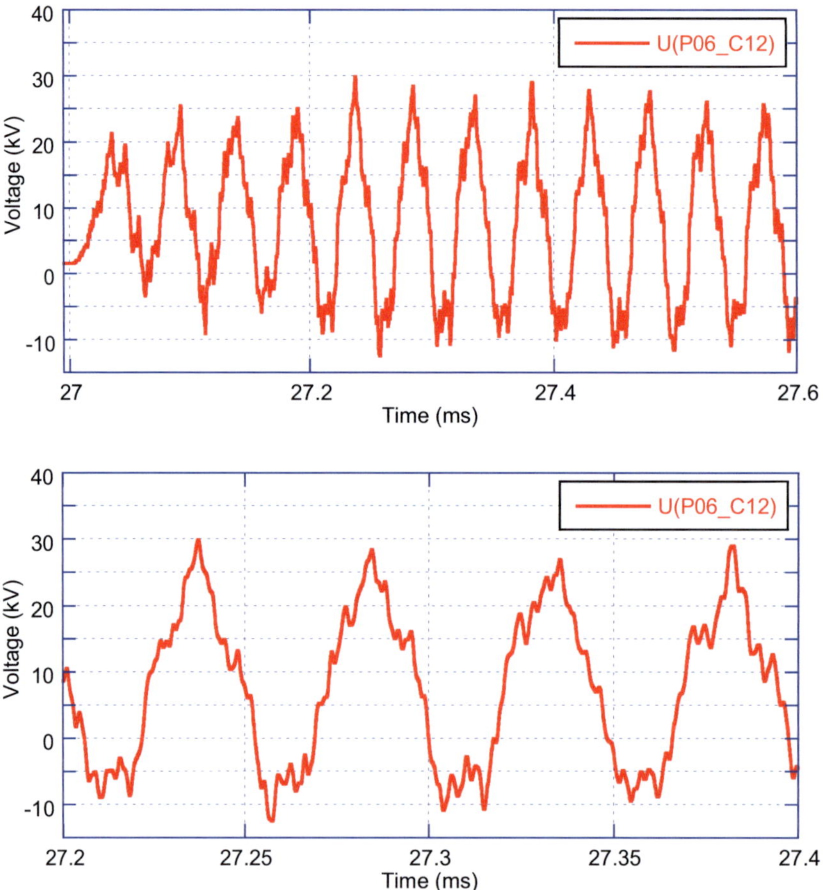

Fig. 6.27: Voltage waveforms of maximum voltage calculated on ground insulation between conductor P06_C12 and ground in failure case 2. Calculated with the 25.5 kHz detailed network model of the PF 3 coil. Maximum time of sampling steps 0.1 µs. Upper part on the time scale between 26.99 ms and 27.60 ms, lower part on the time scale between 27.2 ms and 27.4 ms.

The rise time of the maximum voltage at the positive terminal of the PF 6 coil was calculated to have a peak-to-peak value of 10.4 µs and corresponded to the rise time of sine voltage at a frequency of about 32 kHz and the calculated resonance frequency of the PF 6 coil. The rise times of internal maximum voltages varied from 14 µs to 17.4 µs depending on the location within the coil.

Tab. 6.11: Maximum voltages calculated with the network models of the PF 3 coil in failure case 2. Maximum voltages calculated on particular insulations are shown in red.

		Terminal to ground	Insulation turn to ground	Insulation between layers	Insulation between turns
0.51 kHz network model	Value of maximum voltage	25.4 kV	29.5 kV	13.9 kV	9.9 kV
	Location of maximum voltage	Positive terminal	P01_C06 to ground	P01_C08 to P02_C08	P01_C06 to P01_C07
	Time of maximum voltage	27.031 ms	27.031 ms	27.177 ms	27.188 ms
	Rise time of maximum voltage	15.8 µs	4.2 µs	3.6 µs	4.4 µs
	Start rise voltage	7.0 kV	13.1 kV	-3.8 kV	-2.5 kV
25.5 kHz network model	Value of maximum voltage	25.4 kV	29.8 kV	17.1 kV	12.3 kV
	Location of maximum voltage	Positive terminal	P06_C12 to ground	P14_C01 to P15_C01	P02_C01 to P02_C02
	Time of maximum voltage	27.031 ms	27.237 ms	27.380 ms	27.310 ms
	Rise time of maximum voltage	15.8 µs	16.5 µs	13.4 µs	4.3 µs
	Start rise voltage	7.0 kV	-8.7 kV	-8.7 kV	-3.0 kV
200 kHz network model	Value of maximum voltage	25.4 kV	27.5 kV	11.4 kV	-10.5 kV
	Location of maximum voltage	Positive terminal	P01_C06 to ground	P15_C06 to P16_C06	P02_C04 to P02_C05
	Time of maximum voltage	27.031 ms	27.086 ms	27.751 ms	27.110 ms
	Rise time of maximum voltage	15.8 µs	5.3 µs	1.6 µs	1.5 µs
	Start rise voltage	7.0 kV	14.4 kV	-1.2 kV	0.7 kV

Tab. 6.12: Maximum voltages calculated with the network models of the PF 6 coil in failure case 2. Maximum voltages calculated on particular insulations are shown in red.

		Terminal to ground	Insulation turn to ground	Insulation between layers	Insulation between turns
0.55 kHz network model	Value of maximum voltage	23.4 kV	25.9 kV	12.0 kV	8.9 kV
	Location of maximum voltage	Positive terminal	P01_C01 to ground	P02_C12 to P03_C12	P16_C12 to P16_C13
	Time of maximum voltage	27.025 ms	27.029 ms	27.058 ms	27.445 ms
	Rise time of maximum voltage	10.4 µs	15.6 µs	11.9 µs	2.3 µs
	Start rise voltage	7.2 kV	7.9 kV	-6.7 kV	-1.5 kV
32 kHz network model	Value of maximum voltage	23.4 kV	30.1 kV	13.1 kV	9.0 kV
	Location of maximum voltage	Positive terminal	P01_C15 to ground	P14_C02 to P15_C02	P16_C18 to P16_C19
	Time of maximum voltage	27.025 ms	27.024 ms	27.207 ms	27.245 ms
	Rise time of maximum voltage	10.4 µs	1.5 µs	17.1 µs	1.9 µs
	Start rise voltage	7.2 kV	14.5 kV	-5.4 kV	-1.1 kV
200 kHz network model	Value of maximum voltage	23.4 kV	27.7 kV	11.6 kV	8.0 kV
	Location of maximum voltage	Positive terminal	P01_C15 to ground	P14_C02 to P15_C02	P16_C16 to P16_C17
	Time of maximum voltage	27.025 ms	27.023 ms	27.135 ms	27.287 ms
	Rise time of maximum voltage	10.4 µs	1.4 µs	17.4 µs	2.1 µs
	Start rise voltage	7.2 kV	15.8 kV	-5.0 kV	0.1 kV

6.4　Summary of Calculation Results in Time and Frequency Domain

The transient electrical behaviour of the PF 3 and PF 6 coils was analysed with detailed network models in frequency and time domains. First, the resonance frequency of both coils was calculated in the frequency domain and gave the first benchmark regarding the transient behaviour of the coils. Due to the frequency-dependent inductances of the coil conductor, an iterative calculation loop was used for calculation of the final resonance frequency. Additionally, the influence of the instrumentation cables and unsymmetrical grounding of the coils on the resonance frequency was analysed with the generated detailed network models of the PF 3 and PF 6 coils. The results of the calculations in frequency domain are summarised in Tab. 6.13. The unsymmetrical grounding and instrumentation cables displaced the resonance frequencies of the PF 3 and PF 6 coils to lower values.

Due to the 50 kΩ current limiting resistor which is planed to be used in at the beginning of ITER instrumentation cables [CDA3], detailed network models without instrumentation cables were used for the calculations of internal voltage distribution within both coils in time domain.

Tab. 6.13: Calculated resonance frequencies of the PF 3 and PF 6 coils and its dependence on unsymmetrical grounding and instrumentation cables.

	PF 3	PF 6
Symmetrical grounding without instrumentation cables	25.5 kHz	32.0 kHz
Symmetrical grounding with instrumentation cables	24.4 kHz	30.7 kHz
Unsymmetrical grounding without instrumentation cables	20.7 kHz	28.6 kHz
Unsymmetrical grounding with instrumentation cables	17.0 kHz	22.9 kHz

The transient behaviour of the PF 3 and PF 6 coils in the time domain was calculated first with the network model of the power supply circuits of the CS PF coil system, where all CS and PF coils had been implemented as simplified inductances calculated with the FEM model of the CS PF coil system. Mutual inductances between each coil calculated with the same FEM model were implemented in the network model as a coupling matrix. The voltage waveforms on the terminals of PF 3 and PF 6 coils were calculated with the network model of the CS PF coil system for four calculation scenarios: the reference scenario, fast discharge and two failure cases. Maximum voltages calculated at four scenarios on terminals of PF 3 and PF 6 coils are summarised in Tab. 6.14. The ITER maximum terminal voltages given for an earth fault on a coil terminal during fast discharge in addition to no-load voltages of AC/DC converters [Lib08] are lower than the calculated maximum voltages in failure case 2, described in section 6.2.4.

Tab. 6.14: Absolute values of maximum voltages at the terminals of the PF 3 and PF 6 coils calculated with the network model of the power supply circuits of CS and PF coils.

	PF 3	PF 6
Reference scenario	14.9 kV	7.2 kV
Fast discharge	5.3 kV	5.0 kV
Failure case 1	19.3 kV	8.5 kV
Failure case 2	25.4 kV	23.4 kV
ITER maximum terminal voltages [Lib08]	23.2 kV	18.9 kV

The voltage waveforms calculated at the terminals of the PF 3 and PF 6 coils with the network model of the power supply circuits of the CS and PF coils were used as excitations of the detailed network models of both coils for calculation of internal voltage distribution. The voltage excitations consisted of oscillations with different frequencies in several frequency ranges up to 300 kHz. The inductances of the PF coils were frequency-dependent and decreased with increasing frequency. The influence of the frequency dependence of inductances on the internal voltage distribution was analysed by use of detailed network models for low, resonance and high frequencies of the PF 3 and PF 6 coils, respectively. Thus, the internal voltage distribution was calculated for four calculation scenarios with detailed network models of the PF 3 and PF 6 coils for low, resonance and high frequencies. Maximum voltages within both coils were calculated on the ground, layer and turn insulations and are summarised in Tab. 6.15. The ITER maximum turn voltage was given only for normal operation [DDD06k] and is comparable with calculated internal voltages in the reference scenario, shown in section 6.3.1. Much higher internal voltages were calculated in fault conditions within PF 3 and PF 6 coils, described in section 6.3.4.

The internal voltage distribution within the PF 3 and PF 6 coils depended on the amplitude and rise time of voltage excitation at the terminals of the coils. Relatively slow excitations during the reference scenario led to a linear voltage distribution and equal voltages were calculated with the detailed network models for low, resonance and high frequencies of both coils.

Fast excitations at the coil terminals during fast discharge and both failure cases led to voltage oscillations and non-linear voltage distribution within the PF 3 and PF 6 coils. The calculated amplitudes of the internal voltages on layer and turn insulation in these scenarios were much higher than in the reference scenario. Additionally, the maximum voltage on the ground insulation within both coils during fast discharge and both failure cases was higher than the maximum calculated voltage at the coil terminals.

Tab. 6.15: Absolute values of maximum internal voltages on the ground, layer and turn insulation calculated with the detailed network models of the PF 3 and PF 6 coils.

	PF 3	PF 6
Terminal voltage in failure case 2	25.4 kV	23.4 kV
Ground insulation in failure case 2	29.8 kV	30.1 kV
Layer insulation in failure case 2	17.1 kV	13.1 kV
Turn insulation in failure case 2	12.3 kV	9.0 kV
Turn insulation in reference scenario	0.9 kV	0.9 kV
ITER maximum turn voltage for normal operation only [DDD06k]	1.0 kV	0.8 kV

7 Summary

The superconducting coil system of ITER will be used for confinement and control of the plasma during the fusion reaction. Verification of the high-voltage insulation co-ordination is essential for a reliable operation of the coils. Transient electrical oscillations occur at the terminals of the superconducting coils due to voltage excitations with high amplitudes and fast rise times. These excitations may lead to internal oscillations and non-linear voltage distribution within the coil. Calculations of the transient electrical behaviour of the coils provide a basis for high-voltage insulation co-ordination and definition of test voltages and waveforms.

The main objective of this work was to calculate the transient electrical behaviour of ITER poloidal field (PF) coils using PF 3 and PF 6 coils as examples. The PF 3 coil was chosen for analysis of internal voltage distribution, because this PF coil has the largest coil diameter. The PF 6 coil was selected, because this PF coil has the highest number of turns. This means, that the highest transient voltages can be expected for both coils. A special strategy was defined to calculate the internal voltage distribution within PF 3 and PF 6 coils. The calculation strategy was divided into three main parts: calculations with the finite element method (FEM) models, calculations with network models in the frequency domain and calculations with network models in the time domain.

The self and mutual inductances of all central solenoid (CS) and PF coils were calculated with a simplified FEM model of the CS PF coil system. The self inductances of PF 3 and PF 6 coils calculated at DC with the FEM model of the CS PF coil system are shown in Tab. 7.1. The frequency-dependent inductances of each turn of the PF 3 and PF 6 coils were calculated with detailed FEM models of both coils. A strong frequency dependence of inductance was calculated with detailed FEM models of the PF 3 and PF 6 coils due to the eddy currents induced in the stainless steel jacket of the coil conductor. The inductances calculated with FEM models were included as lumped elements in the network models of the CS PF coil system and detailed network models of both coils.

The detailed network models of the PF 3 and PF 6 coils were set up with the values of the lumped elements. The capacitances were calculated analytically with the formulas for parallel plate and cylindrical capacitances. The inductances were calculated with detailed FEM models of both coils. First, the detailed network models were used for calculation of resonance frequency of the PF 3 and PF 6 coils in the frequency domain. Due to the frequency dependence of the inductances, an iterative loop was used for calculation of the final resonance frequency with symmetrical grounding of the coils. Additionally, the influence of unsymmetrical grounding and instrumentation cables on the resonance frequency was analysed in the frequency domain. The resonance frequency of the PF 3 and PF 6 coils with symmetrical grounding and without instrumentation cables are shown in Tab. 7.1. Unsymmetrical grounding and instrumentation cables displace the resonance frequency of the coils to lower values.

The network calculations in the time domain were started by calculation of voltage waveforms at the terminals of the PF 3 and PF 6 coils with the network model of the CS PF coil system. The voltage waveforms were calculated for four specified calculation scenarios: the reference scenario, fast discharge and two failure cases. Maximum voltages calculated at the terminals of PF 3 and PF 6 coil and their rise times are shown in Tab. 7.1. These calculated terminal voltages were higher than the ITER maximum terminal voltages described in [Lib08]. Therefore, it is recommended to adjust the test voltages for the ITER PF coils to higher values.

The voltage waveforms calculated with the network model of the CS PF coil system at the terminals of the PF 3 and PF 6 coils were included in detailed network models of both coils as voltage excitations at the coil terminals for calculation of internal voltage distribution. The voltage excitations consisted of oscillations with different frequencies in several frequency ranges up to 300 kHz. The inductances of the PF coils were frequency-dependent and decreased with increasing frequency. The influence of the frequency dependence of inductances on the internal voltage distribution was analysed by using detailed network models for low, resonance and high frequencies of the PF 3 and PF 6 coils, respectively. The internal voltage distribution was calculated for four calculation scenarios with detailed network models of the PF 3 and PF 6 coils for low, resonance and high frequencies. Maximum voltages within both coils were calculated on the ground, layer and turn insulations and are summarised in Tab. 7.1. The ITER specified maximum turn voltages were much lower [DDD06k] and were given for normal operation, only.

Tab. 7.1:   Calculation results of the PF 3 and PF 6 coils and its comparison with ITER specified maximum voltages.

	PF 3	PF 6
FEM calculated DC self inductance	1.83 H	2.02 H
Resonance frequency	25.5 kHz	32.0 kHz
Maximum voltages on coil terminals	25.4 kV	23.4 kV
Rise time of maximum terminal voltage	15.8 μs	10.4 μs
ITER maximum terminal voltage [Lib08]	23.2 kV	18.9 kV
Maximum voltage on ground insulation	29.8 kV	30.1 kV
Maximum voltage on layer insulation	17.1 kV	13.1 kV
Maximum voltage on turn insulation	12.3 kV	9.0 kV
ITER maximum turn voltage at normal operation only [DDD06k]	1.0 kV	0.8 kV

The internal voltage distribution within the PF 3 and PF 6 coils depended on the amplitude and rise time of voltage excitation at the terminals of the coils. Relatively slow excitations during the reference scenario led to a linear voltage distribution and equal voltage values calculated with the detailed network models for low, resonance and high frequencies of both coils. Fast excitations at the coil terminals during fast discharge and both failure cases led to voltage oscillations and non-linear voltage distribution within the PF 3 and PF 6 coils. Thus, the maximum voltage on the ground insulation within both coils was calculated higher than the voltage at the coil terminals.

A Annex

A.1   Detailed Data of FEM Calculations

FEM Model of CS PF Coil System

Model Definition

Solver: Magnetostatic   Symmetry: RZ-Plane

Drawing Size: Z-direction: ±30 m; R-direction: 40 m

Tab. A.2: Dimensions of the FEM model of CS PF coil system for Maxwell 2D.

Coil name	Location of starting point		Model dimensions	
	R-direction	Z-direction	R-direction	Z-direction
CS3U	1.360 m	4.300 m	0.719 m	2.091 m
CS2U	1.360 m	2.160 m	0.719 m	2.091 m
CS1U	1.360 m	0.025 m	0.719 m	2.091 m
CS1L	1.360 m	-2.110 m	0.719 m	2.091 m
CS2L	1.360 m	-4.250 m	0.719 m	2.091 m
CS3L	1.360 m	-6.390 m	0.719 m	2.091 m
PF 1	3.459 m	7.069 m	0.968 m	0.976 m
PF 2	7.995 m	6.233 m	0.649 m	0.595 m
PF 3	11.643 m	2.782 m	0.708 m	0.966 m
PF 4	11.643 m	-2.726 m	0.649 m	0.966 m
PF 5	7.985 m	-7.203 m	0.820 m	0.945 m
PF 6	3.447 m	-8.045 m	1.633 m	0.976 m

Materials

CS and PF coils: Copper with relative permeability = 1

Background: Vacuum with relative permeability = 1

Setup Boundaries and Sources

Current Source	All superconducting cables are stranded with 45 kA
Boundary	Balloon

Setup Executive Parameters

All coils were chosen into the impedance matrix.
Return path of the current was set to default.

Setup Solution Options

Mesh: Initial mesh
Percent refinement per pass: 5
Number of requested passes: 70
Percent error: 0.1

Convergence Data

Number of complete passes	Number of triangles	Energy Error (%)
70	62561	0.1895

Tab. A.3: Impedance matrix calculated with simplified FEM model of CS PF coil system.

	CS3U	CS2U	CS1U	CS1L	CS2L	CS3L
CS3U	2.58E-06 H					
CS2U	8.00E-07 H	2.58E-06 H				
CS1U	1.74E-07 H	8.00E-07 H	2.58E-06 H			
CS1L	5.94E-08 H	1.74E-07 H	8.00E-07 H	2.58E-06 H		
CS2L	2.64E-08 H	5.94E-08 H	1.74E-07 H	8.00E-07 H	2.58E-06 H	
CS3L	1.39E-08 H	2.64E-08 H	5.94E-08 H	1.74E-07 H	8.00E-07 H	2.58E-06 H
PF 1	9.97E-07 H	4.43E-07 H	2.05E-07 H	1.06E-07 H	6.03E-08 H	3.70E-08 H
PF 2	6.98E-07 H	5.73E-07 H	4.15E-07 H	2.85E-07 H	1.93E-07 H	1.33E-07 H
PF 3	4.73E-07 H	4.95E-07 H	4.71E-07 H	4.11E-07 H	3.36E-07 H	2.63E-07 H
PF 4	2.97E-07 H	3.73E-07 H	4.44E-07 H	4.89E-07 H	4.92E-07 H	4.49E-07 H
PF 5	1.30E-07 H	1.88E-07 H	2.76E-07 H	4.01E-07 H	5.55E-07 H	6.83E-07 H
PF 6	4.25E-08 H	6.85E-08 H	1.19E-07 H	2.24E-07 H	4.63E-07 H	9.72E-07 H

Tab. A.4: Impedance matrix calculated with simplified FEM model of CS PF coil system.

	PF 1	PF 2	PF 3	PF 4	PF 5	PF 6
PF 1	1.14E-05 H					
PF 2	3.94E-06 H	3.64E-05 H				
PF 3	2.17E-06 H	1.12E-05 H	5.34E-05 H			
PF 4	1.16E-06 H	5.26E-06 H	1.43E-05 H	5.38E-05 H		
PF 5	4.48E-07 H	1.89E-06 H	4.53E-06 H	9.76E-06 H	3.31E-05 H	
PF 6	1.33E-07 H	5.30E-07 H	1.19E-06 H	2.32E-06 H	4.74E-06 H	1.12E-05 H

Detailed FEM Model of PF 3 coil

FEM Model of the PF 3 Coil for DC

Model Definition

Solver: Magnetostatic   Symmetry: RZ-Plane

Drawing Size: Z-direction: ±10 m; R-direction: 20 m

Geometry:

Cooling Tube P01_C01:   Circle centre: 11672.25 mm; 452.75 mm; Radius: 6 mm

Superconducting cable P01_C01:   Circle centre: 11672.25 mm; 452.75 mm; Radius: 17.2 mm

Jacket P01_C01:   First corner: 11646.2 mm; 426.7 mm

                               Second corner: 11698.3 mm; 478.8 mm

Copy of the conductor P01_C01 downwards one time with a distance of 59.5 mm

Copy of the conductor P02_C01 downwards one time with a distance of 61.5 mm with consideration of additionally insulation between the double pancakes of 2 mm

Continuing of copy of the conductors downwards with distance depending on the location in the double pancake

Copy of the stack of the 16 first conductors rightwards 12 times with a distance of 59 mm

Materials

Cooling tube   Vacuum, relative permittivity = 1, relative permeability = 1

Superconducting cable   Copper at 4K, relative permittivity = 1, relative permeability = 1, conductivity = 6.4E9 S/m

Stainless steel jacket   Stainless steel at 4K, relative permittivity = 1,
          relative permeability = 1, conductivity = 1.88e6 S/m

Insulation   Epoxy resin, relative permittivity = 4, relative permeability = 1

Background: Vacuum, relative permittivity = 1, relative permeability = 1

Setup Boundaries and Sources

Current Source   All superconducting cables are stranded with 45 kA

Boundary   Balloon

Setup Executive Parameters

All superconducting cables were chosen into the impedance matrix.

Return path of the current was set to default.

Setup Solution Options

Mesh: Manual mesh

Object	Insulation	Cooling tube	Superconducting cable	Stainless steel jacket	Background
Number of triangles	52124	70	430	750	67290

FEM Model of PF 3 Coil for Frequencies lower than 5 kHz

Model Definition

Solver: Eddy Current   Symmetry: RZ-Plane
Drawing Size: Z-direction: +10 m; R-direction: 20 m

Materials

Cooling tube   Vacuum, relative permittivity = 1, relative permeability = 1
Superconducting cable   Copper at 4K, relative permittivity = 1, relative permeability = 1,
conductivity = 6.4E9 S/m
Stainless steel jacket   Stainless steel at 4K, relative permittivity = 1,
        relative permeability = 1, conductivity = 1.88e6 S/m
Insulation   Epoxy resin, relative permittivity = 4, relative permeability = 1
Background: Vacuum, relative permittivity = 1, relative permeability = 1

Setup Boundaries and Sources

Current Source   All superconducting cables are stranded with 45 kA
Boundary   Bottom edge of the model boundary = even symmetry
                     Other three edges of the model boundary = balloon

Setup Executive Parameters

All superconducting cables were chosen into the impedance matrix.
Return path of the current was set to default.

Setup Solution Options

Mesh: Manual mesh

Object	Insulation	Cooling tube	Superconducting cable	Stainless steel jacket	Background
Number of triangles	70000	10	48	350	26000

FEM Model of PF 3 Coil for Frequencies higher than 5 kHz

Model Definition

Solver: Eddy Current   Symmetry: RZ-Plane
Drawing Size: Z-direction: ±5 m; R-direction: 10 m

Materials

Cooling tube   Vacuum, relative permittivity = 1, relative permeability = 1
Superconducting cable   Copper at 4K, relative permittivity = 1,
                                relative permeability = 1, conductivity = 6.4E9 S/m
Stainless steel jacket   Stainless steel at 4K, relative permittivity = 1,
                                relative permeability = 1, conductivity = 1.88e6 S/m
Insulation   Epoxy resin, relative permittivity = 4, relative permeability = 1
Background: Vacuum, relative permittivity = 1, relative permeability = 1

Setup Boundaries and Sources

Current Source   All superconducting cables are stranded with 45 kA
Boundary   Balloon

Setup Executive Parameters

All superconducting cables were chosen into the impedance matrix.
Return path of the current was set to default.

Setup Solution Options

Mesh: Manual mesh

Object	Insulation	Cooling tube	Superconducting cable	Stainless steel jacket	Background
Number of triangles	1500	22	150	230	1300

Detailed FEM Model of PF 6 coil

FEM Model of the PF 6 Coil for DC

Model Definition

Solver: Magnetostatic   Symmetry: RZ-Plane

Drawing Size: Z-direction: ±5 m; R-direction: 10 m

Geometry:

Cooling Tube P01_C01:   Circle centre: 3476.25 mm; 458 mm; Radius: 6 mm

Superconducting cable P01_C01:   Circle centre: 3476.25 mm; 458 mm; Radius: 19 mm

Jacket P01_P01:   First corner: 3449.45 mm; 431.2 mm

                                      Second corner: 3503.05 mm; 484.8 mm

Copy of the conductor P01_C01 downwards one time with a distance of 60.7 mm

Copy of the conductor P02_C01 downwards one time with a distance of 60.7 mm with consideration of additionally insulation between the double pancakes of 1 mm

Continuing of copy of the conductors downwards with distance depending on the location in the double pancake

Copy of the stack of the 16 first conductors rightwards 27 times with a distance of 60.5 mm

Materials

Cooling tube   Vacuum, relative permittivity = 1, relative permeability = 1

Superconducting cable   Copper at 4K, relative permittivity = 1, relative permeability = 1,
conductivity = 6.4E9 S/m

Stainless steel jacket Stainless steel at 4K, relative permittivity = 1,
        relative permeability = 1, conductivity = 1.88e6 S/m

Insulation   Epoxy resin, relative permittivity = 4, relative permeability = 1

Background: Vacuum, relative permittivity = 1, relative permeability = 1

Setup Boundaries and Sources

Current Source   All superconducting cables are stranded with 45 kA

Boundary   Balloon

Setup Executive Parameters

All superconducting cables were chosen into the impedance matrix.

Return path of the current was set to default.

Setup Solution Options

Mesh: Manual mesh

Object	Insulation	Cooling tube	Superconducting cable	Stainless steel jacket	Background
Number of triangles	10000	34	72	45	11400

FEM Model of the PF 6 Coil for Frequencies Lower than 5 kHz

Model Definition

Solver: Eddy Current | Symmetry: RZ-Plane
Drawing Size: | Z-direction: +5 m; R-direction: 10 m

Materials

Cooling tube | Vacuum, relative permittivity = 1, relative permeability = 1
Superconducting cable conductivity = 6.4E9 S/m | Copper at 4K, relative permittivity = 1, relative permeability = 1,
Stainless steel jacket | Stainless steel at 4K, relative permittivity = 1, relative permeability = 1, conductivity = 1.88e6 S/m
Insulation | Epoxy resin, relative permittivity = 4, relative permeability = 1
Background: | Vacuum, relative permittivity = 1, relative permeability = 1

Setup Boundaries and Sources

Current Source | All superconducting cables are stranded with 45 kA
Boundary | Balloon

Setup Executive Parameters

All superconducting cables were chosen into the impedance matrix.
Return path of the current was set to default.

Setup Solution Options

Mesh: Manual mesh

Object	Insulation	Cooling tube	Superconducting cable	Stainless steel jacket	Background
Number of triangles	70000	34	72	302	5000

FEM Model of the PF 6 Coil for Frequencies Higher than 5 kHz

Model Definition

Solver: Eddy Current   Symmetry: RZ-Plane

Drawing Size: Z-direction: ±5 m; R-direction: 10 m

Materials

Cooling tube   Vacuum, relative permittivity = 1, relative permeability = 1

Superconducting cable   Copper at 4K, relative permittivity = 1, relative permeability = 1, conductivity = 6.4E9 S/m

Stainless steel jacket   Stainless steel at 4K, relative permittivity = 1, relative permeability = 1, conductivity = 1.88e6 S/m

Insulation   Epoxy resin, relative permittivity = 4, relative permeability = 1

Background: Vacuum, relative permittivity = 1, relative permeability = 1

Setup Boundaries and Sources

Current Source   All superconducting cables are stranded with 45 kA

Boundary   Balloon

Setup Executive Parameters

All superconducting cables were chosen into the impedance matrix.

Return path of the current was set to default.

Setup Solution Options

Mesh: Manual mesh

Object	Insulation	Cooling tube	Superconducting cable	Stainless steel jacket	Background
Number of triangles	3600	34	1200	2600	1200

FEM Models of the Bus Bars for Power Supply

FEM Model of the Superconducting Bus Bars

Model Definition

Solver: Eddy Current   Symmetry: XY-Plane   Frequency 0 Hz

Drawing Size: X-direction: ±100 mm; Y-direction: ±100 mm

Materials

Cooling tube   Vacuum, relative permittivity = 1, relative permeability = 1

Superconducting cable   Copper at 4K, relative permittivity = 1, relative permeability = 1,
conductivity = 6.4E9 S/m

Stainless steel jacket   Stainless steel at 4K, relative permittivity = 1,
                              relative permeability = 1, conductivity = 1.88e6 S/m

Insulation   Epoxy resin, relative permittivity = 4, relative permeability = 1

Background: Vacuum, relative permittivity = 1, relative permeability = 1

Setup Boundaries and Sources

Current Source   Superconducting bus bar with 45 kA

Boundary   Balloon

Setup Executive Parameters

Only superconducting bus bar was chosen into the impedance matrix.
Return path of the current was set to default.

Setup Solution Options

Mesh: Manual mesh

Object	Insulation	Cooling tube	Superconducting cable	Stainless steel jacket	Background
Number of triangles	182	34	1294	73	291

FEM Model of the Water Cooled Bus Bars

Model Definition

Solver: Eddy Current   Symmetry: XY-Plane   Frequency 0 Hz

Drawing Size: X-direction: ±400 mm; Y-direction: ±400 mm

Materials

Water cooling tube   Vacuum, relative permittivity = 1, relative permeability = 1

Aluminium bus bar   Aluminium, relative permittivity = 1, relative permeability = 1,
conductivity = 3.6e+7

Background: Vacuum, relative permittivity = 1, relative permeability = 1

Setup Boundaries and Sources

Current Source   Water cooled bus bar with 45 kA

Boundary   Balloon

Setup Executive Parameters

Water cooled bus bar was chosen into the impedance matrix.

Return path of the current was set to default.

Setup Solution Options

Mesh: Manual mesh

Object	Water cooling tube	Water cooled bus bar	background
Number of triangles	34	1628	672

A.2   Frequency Behaviour of the Resistance of NbTi Coils

The high-voltage insulation co-ordination of the ITER coils requires calculation of the transient electrical behaviour of these coils for different operation scenarios and failure cases to define voltage amplitudes and procedures for testing the different insulation types. For accurate modelling and implementation of the superconducting cables in network models of the ITER PF coils which will be made of an NbTi superconductor, it is important to know the resistance of superconducting coils at voltage excitations in the relevant frequency range up to 300 kHz. The high-frequency excitations of PF coils will be mainly voltage excitations, e. g. during the fast discharge up to 5 kV [PAF6Bd], with relatively low AC currents. Measurements of the resistances of two small-sized NbTi coils were performed to analyse the behaviour of the resistance of these coils at frequencies up to 1 MHz for voltage excitations with low currents. This chapter describes the NbTi coils analysed in the experiments, the laboratory equipment which was used and the results of the measurements.

NbTi Superconducting Coils

The measurements were made with two NbTi coils shown in Fig. A.1. Two turns were NbTi copper-free turns in the centre of each coil. The remaining three turns were copper-stabilised NbTi superconducting windings. The total number of turns was 5 for each coil. Solid copper rings were installed at the top and bottom of the coils to ensure stable mechanical connection to the current leads and low electrical resistance. The height of the coils without solid copper rings was 50 mm, with solid copper rings 80 mm. The first coil had 24 superconducting filaments (F24), the second coil contained 54 superconducting filaments (F54). The outer dimensions were the same for both coils, the only difference was in the number of filaments and the corresponding properties of the NbTi conductors especially at room temperature. The detailed data of the F24 and F54 coils are listed in Tab. A.5.

Both coils had two pairs of wires for voltage measurement, the first one across the whole coil with 5 turns and the second coil installed across the two copper-free NbTi turns in the centre of the coil. The measuring wires were equipped with inductance compensation to measure the real part of the voltage across the coil without its imaginary part.

The NbTi coils were placed in a helium cryostat, a so-called Jumbo facility [Rim00], in the high-field lab of the Institute of Technical Physics (ITEP) of Karlsruhe Institute of Technology (KIT). The cryostat was equipped with superconducting coils inside for background magnetic field experiments. The background field coils can supply DC magnetic peak fields up to 10.5 T in a bore of 100 mm. All coils were cooled with liquid helium at a temperature of 4.2 K.

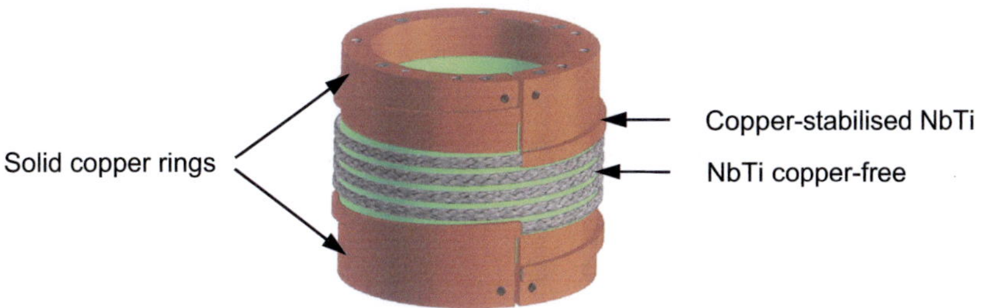

Fig. A.1: Schematic view of the NbTi coil used in the experiments. The solid copper rings at the top and bottom of the coil were used for connection with current leads. Two NbTi copper-free turns are located in the centre of the coil. Three copper-stabilised NbTi turns are arranged equally at the top and bottom of the coil.

Tab. A.5: Detailed data of the F24 and F54 NbTi coils.

	F24 NbTi coil	F54 NbTi coil
Number of superconducting filaments per strand	24	54
Copper to NbTi ratio in copper-stabilised turns	6.8	1.3
Cumulative cross-section of NbTi filaments	0.049 mm²	0.168 mm²
Diameter of one filament	51 μm	63 μm
Strand diameter	0.7 mm	0.7 mm
Number of turns	5	5
Number of turns without copper stabilisation	2	2
Diameter of the coil	90 mm	90 mm
Height of the coil	50 mm	50 mm
Height of the coil with solid copper rings	80 mm	80 mm
Inductance of each turn	0.44 μH	0.44 μH
Capacitance between two turns	0.78 pF	0.78 pF
Capacitance of one turn to ground	1.2 pF	1.2 pF

The NbTi coils used in the experiments were connected to current leads of 1.4 m in length between the cryostat flange and the coil. The coil with current leads in front of the cryostat is shown in Fig. A.2. The current leads have two terminals outside of the cryostat for connection to a power supply. The coils were excited with high-frequency voltages using a frequency generator connected to the coil terminals outside of the cryostat. The sample holder was placed between both current leads. The measurement wires were arranged inside the sample holder. A plug at the top of the sample holder outside of the cryostat was connected to an analysis box. The analysis box was connected to an oscilloscope.

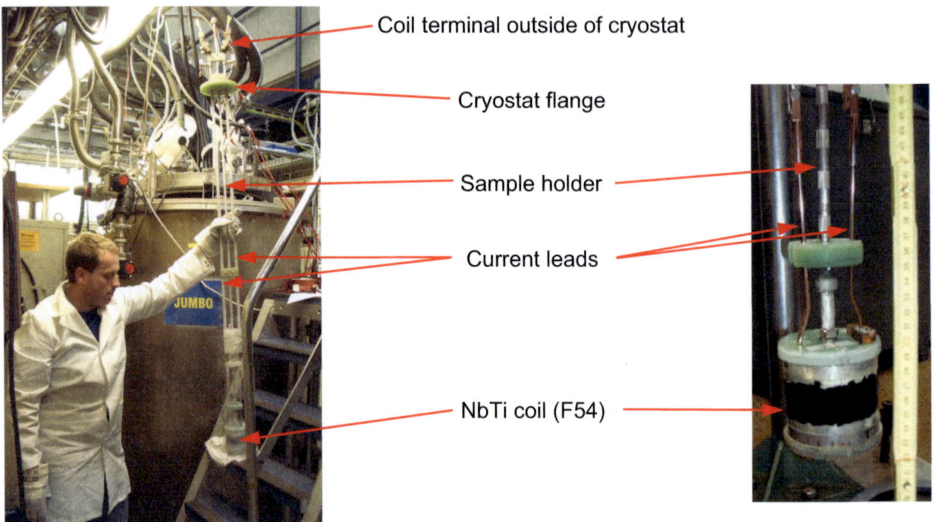

Fig. A.2: The left figure shows the NbTi coil with current leads, sample holder, cryostat flange and coil terminals in front of the cryostat in the high-field lab of ITEP. A closer view on NbTi coil with current leads and sample holder is given on the right.

Laboratory Set-up and Measurement Procedure

The frequency behaviour of the resistance of the NbTi coils was analysed with a measuring circuit consisting of frequency generator, shunt resistor and oscilloscope. The measurements were carried out at frequencies of up to 1 MHz. The laboratory set-up is shown schematically in Fig. A.3.

The voltage excitations were made using a frequency generator with maximum voltages of about 8 V depending on the impedance of the circuit. The current was measured with a shunt resistor of 2.9 Ω. According to analysis done before the experiments, the frequency-dependence of the shunt resistor was neglectable. Five voltages were measured with an oscilloscope during the experiments. The voltage of the shunt resistor was used to calculate the current of the coil circuit. The current in the circuit was about 100 mA, which is very low compared to the critical current of the NbTi F24 coil of about 26 A measured at 9 T background magnetic field. The critical current of F54 is even higher, 96 A at 9 T.

The voltages were additionally measured at the frequency generator, coil terminals, the five turns of the coil and the two NbTi copper-free turns. The resistance of each part of the circuit was calculated with the current, the appropriate voltage in this part of the circuit and the phase angle between current and voltage, e.g. the resistance of two NbTi copper-free turns was calculated with the voltage of these turns and the current through the circuit.

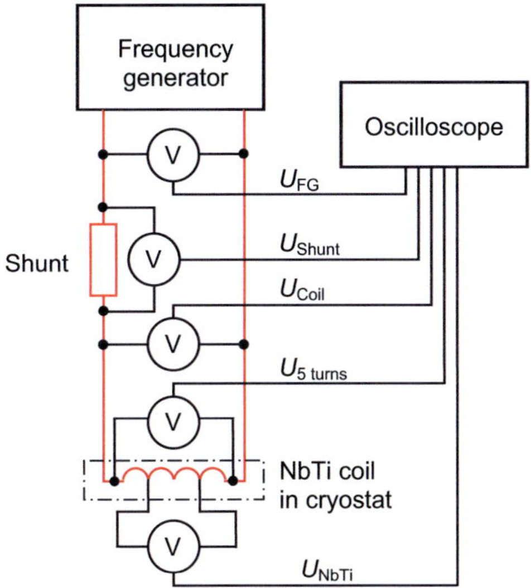

Fig. A.3: Schematic representation of the laboratory set-up for measurement of the frequency behaviour of the resistance of NbTi coils with a frequency generator. The voltages of the frequency generator, shunt resistor, coil terminals, 5 turns of the coil and two NbTi copper-free turns of the coil were measured with an oscilloscope.

The measurements were performed at room temperature, at 77 K and at 4.2 K with 0 T background magnetic field. The measurements at 4.2 K were additionally made with 4 T, 8 T, and 10.5 T background DC magnetic fields. The measurement results of the experiments are shown in the next section.

Measurement Results

The main objective of the measurements was to analyse the frequency behaviour of the resistance of the NbTi coils in the superconducting state. The relative differences between the resistances of NbTi at room temperature, 77 K and 4.2 K and its frequency behaviour were of particular importance. These results were used for computer-based modelling of the ITER PF coils. The frequency behaviour of the F24 and F54 NbTi coils was measured at different temperatures and different background magnetic DC fields. The results are shown in Fig. A.4 and Fig. A.5.

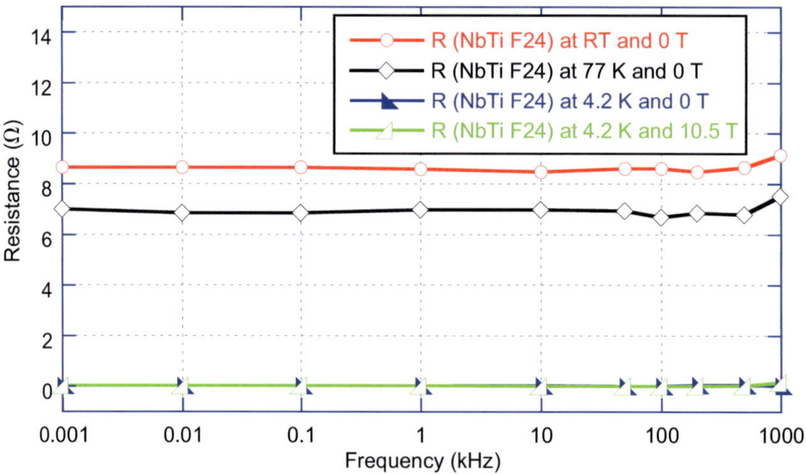

Fig. A.4:   Frequency dependence of the resistance of the NbTi coil with 24 superconducting filaments (F24) at different temperatures and DC background magnetic fields.

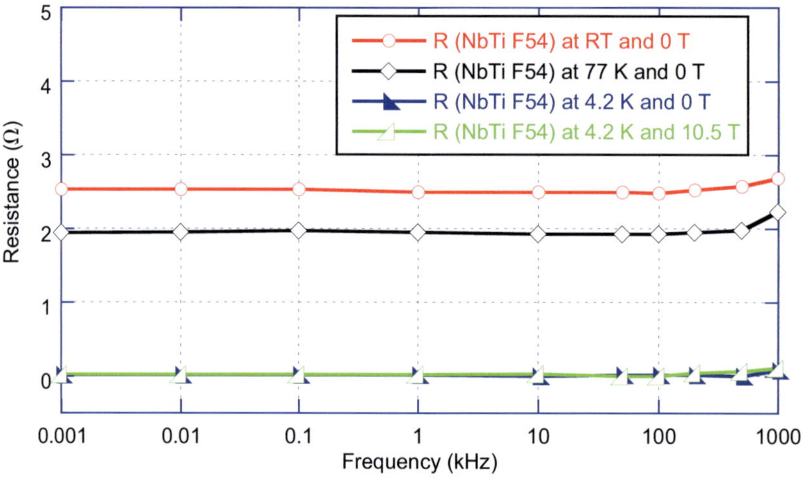

Fig. A.5:   Frequency dependence of the resistance of the NbTi coil with 54 superconducting filaments (F54) at different temperatures and DC background magnetic fields.

The resistances of the F24 and F54 coils at room temperature (RT) and at 77 K were constant in the frequency range up to 500 kHz with a scatter of 2 %. The resistivity of the NbTi F24 coil at RT was calculated to be 0.72 $\mu\Omega$m, for F54 it was 0.73 $\mu\Omega$m. Both values were in the range of the resistivity of NbTi alloys at RT having various Nb to Ti ratios [Feh77]. At room temperature and

77 K, the resistance of the F24 and F54 coils increased slightly at 1 MHz. The influence of capacitances in these measurements was neglected due to their relatively low values between two turns and between one turn and ground, as shown in Tab. A.5. Even at a frequency of 1 MHz, the inductive reactance of one turn was much lower, $X_L = 2.8\ \Omega$, than the capacitive reactance of one turn to ground and between two turns of $X_{CS} = 204\ k\Omega$ and $X_{CD} = 132\ k\Omega$, respectively.

At 4.2 K, the resistance of both coils was constant at a value of about 20 mΩ over the whole frequency range with a scatter of 10 mΩ. The resolution of the measurement was 10 mΩ. Hence, the resistance of the coils measured in the superconducting state is the resistance of the measuring cables and cable connections.

The background DC magnetic field reduced the critical current of the NbTi conductor, especially with excitations at high frequency. Before the experiment, it had been expected that the NbTi coil would show an increase of resistance at excitations with high frequencies of up to 1 MHz in addition to a high magnetic background field. In experiments with a background DC magnetic field of up to 10.5 T at 4.2 K, frequency dependence of the resistances of the F24 and F54 coils was not measurable. The graphs of the resistances of the F24 and F54 coils at 4.2 K with background DC magnetic fields of 4 T and 8 T are not shown, but had similar values and similar frequency behaviour than both coils at 0 T and 10.5 T.

Conclusion

Calculation of the electrical behaviour of ITER PF coils requires accurate coil models. For the implementation of the superconducting turns in the network models of the coils, it is important to know the frequency behaviour of the resistance of the superconducting cables. The resistance of two NbTi superconducting coils was measured in the relevant frequency range of up to 1 MHz. Additionally, the influence of the background magnetic DC field of up to 10.5 T on the frequency dependence of the resistance of NbTi coils was analysed.

It was concluded from these measurements that the NbTi coils remained superconducting at frequencies of up to 1 MHz with voltage excitations and low currents. The very low resistance was also measured at a background magnetic field of up to 10.5 T in the whole frequency range.

As a consequence, the superconducting cables of the ITER PF coils made of NbTi superconductor material were implemented as negligible resistances, e. g. 100 pΩ, in the network models of the PF coils.

A.3   Detailed Data for Network Models

Network Model of CS PF Coil System

Tab. A.6: Capacitances used in network model of the CS PF coil system calculated as a sum of capacitances between turns within the coils and from turns to ground.

Name of the coil	Capacitance within the coil	Capacitance to ground	Name of the coil	Capacitance within the coil	Capacitance to ground
CS3U	1.42E-09 F	11.4E-09 F	PF 1	1.47E-09 F	11.3E-09 F
CS2U	1.42E-09 F	11.4E-09 F	PF 2	3.84E-09 F	21.5E-09 F
CS1U	1.42E-09 F	11.4E-09 F	PF 3	4.03E-09 F	30.7E-09 F
CS1L	1.42E-09 F	11.4E-09 F	PF 4	4.12E-09 F	31.5E-09 F
CS2L	1.42E-09 F	11.4E-09 F	PF 5	1.86E-09 F	14.5E-09 F
CS3L	1.42E-09 F	11.4E-09 F	PF 6	1.78E-09 F	13.4E-09 F

Tab. A.7: Capacitances, inductances and resistances used in the network model of the CS PF coil system for water cooled bus bars (wcbb).

Name of the element	Value of the element	Name of the element	Value of the element	Name of the element	Value of the element
Cwcbb+CS1U	4.00E-10 F	Lwcbb+CS1U	4.20E-05 H	Rwcbb+CS1U	4.20E-05 Ω
Cwcbb+CS2L	4.00E-10 F	Lwcbb+CS2L	4.17E-05 H	Rwcbb+CS2L	4.17E-05 Ω
Cwcbb+CS2U	4.00E-10 F	Lwcbb+CS2U	4.17E-05 H	Rwcbb+CS2U	4.17E-05 Ω
Cwcbb+CS3L	4.00E-10 F	Lwcbb+CS3L	4.17E-05 H	Rwcbb+CS3L	4.17E-05 Ω
Cwcbb+CS3U	3.90E-10 F	Lwcbb+CS3U	4.17E-05 H	Rwcbb+CS3U	4.17E-05 Ω
Cwcbb+PF1	3.90E-10 F	Lwcbb+PF1	4.17E-05 H	Rwcbb+PF1	4.17E-05 Ω
Cwcbb+PF2	4.00E-10 F	Lwcbb+PF2	4.20E-05 H	Rwcbb+PF2	4.20E-05 Ω
C wcbb+PF3	4.00E-10 F	Lwcbb+PF3	4.20E-05 H	Rwcbb+PF3	4.20E-05 Ω
Cwcbb+PF4	4.00E-10 F	Lwcbb+PF4	4.20E-05 H	Rwcbb+PF4	4.20E-05 Ω
Cwcbb+PF5	4.00E-10 F	Lwcbb+PF5	4.20E-05 H	Rwcbb+PF5	4.20E-05 Ω
Cwcbb+PF6	3.90E-10 F	Lwcbb+PF6	4.17E-05 H	Rwcbb+PF6	4.17E-05 Ω

Cwcbb-1CS1L	4.00E-10 F	Lwcbb-1CS1L	4.20E-05 H	Rwcbb-1CS1L	4.20E-05 Ω
Cwcbb-1CS2L	4.00E-10 F	Lwcbb-1CS2L	4.17E-05 H	Rwcbb-1CS2L	4.17E-05 Ω
Cwcbb-1CS2U	4.00E-10 F	Lwcbb-1CS2U	4.17E-05 H	Rwcbb-1CS2U	4.17E-05 Ω
Cwcbb-1CS3L	4.00E-10 F	Lwcbb-1CS3L	4.17E-05 H	Rwcbb-1CS3L	4.17E-05 Ω
Cwcbb-1CS3U	3.90E-10 F	Lwcbb-1CS3U	4.17E-05 H	Rwcbb-1CS3U	4.17E-05 Ω
Cwcbb-1PF1	3.90E-10 F	Lwcbb-1PF1	4.17E-05 H	Rwcbb-1PF1	4.17E-05 Ω
Cwcbb-1PF2	4.00E-10 F	Lwcbb-1PF2	4.20E-05 H	Rwcbb-1PF2	4.20E-05 Ω
Cwcbb-1PF3	4.00E-10 F	Lwcbb-1PF3	4.20E-05 H	Rwcbb-1PF3	4.20E-05 Ω
Cwcbb-1PF4	4.00E-10 F	Lwcbb-1PF4	4.20E-05 H	Rwcbb-1PF4	4.20E-05 Ω
Cwcbb-1PF5	4.00E-10 F	Lwcbb-1PF5	4.20E-05 H	Rwcbb-1PF5	4.20E-05 Ω
Cwcbb-1PF6	3.90E-10 F	Lwcbb-1PF6	4.17E-05 H	Rwcbb-1PF6	4.17E-05 Ω
Cwcbb-2CS1L	5.90E-10 F	Lwcbb-2CS1L	6.30E-05 H	Rwcbb-2CS1L	6.30E-05 Ω
Cwcbb-2CS2L	5.90E-10 F	Lwcbb-2CS2L	6.30E-05 H	Rwcbb-2CS2L	6.30E-05 Ω
Cwcbb-2CS2U	4.00E-10 F	Lwcbb-2CS2U	4.33E-05 H	Rwcbb-2CS2U	4.33E-05 Ω
Cwcbb-2CS3L	5.90E-10 F	Lwcbb-2CS3L	6.30E-05 H	Rwcbb-2CS3L	6.30E-05 Ω
Cwcbb-2CS3U	4.00E-10 F	Lwcbb-2CS3U	4.33E-05 H	Rwcbb-2CS3U	4.33E-05 Ω
Cwcbb-2PF1	4.00E-10 F	Lwcbb-2PF1	4.33E-05 H	Rwcbb-2PF1	4.33E-05 Ω
Cwcbb-2PF2	4.00E-10 F	Lwcbb-2PF2	4.30E-05 H	Rwcbb-2PF2	4.30E-05 Ω
Cwcbb-2PF3	4.00E-10 F	Lwcbb-2PF3	4.30E-05 H	Rwcbb-2PF3	4.30E-05 Ω
Cwcbb-2PF4	5.90E-10 F	Lwcbb-2PF4	6.30E-05 H	Rwcbb-2PF4	6.30E-05 Ω
Cwcbb-2PF5	5.90E-10 F	Lwcbb-2PF5	6.30E-05 H	Rwcbb-2PF5	6.30E-05 Ω
Cwcbb-2PF6	4.00E-10 F	Lwcbb-2PF6	4.33E-05 H	Rwcbb-2PF6	4.33E-05 Ω
CwcbbCS1U	1.05E-09 F				
CwcbbCS2L	1.05E-09 F				
CwcbbCS2U	1.05E-09 F				
CwcbbCS3L	1.05E-09 F				
CwcbbCS3U	1.05E-09 F				
CwcbbPF1	1.05E-09 F				
CwcbbPF2	1.05E-09 F				
CwcbbPF3	1.05E-09 F				
CwcbbPF4	1.05E-09 F				
CwcbbPF5	1.05E-09 F				
CwcbbPF6	1.05E-09 F				

Tab. A.8: Capacitances, inductances and resistances used in the network model of the CS PF coil system for connection of switching network units (SNU) with switching network resistors (SNR) by six parallel power cables.

Name of the element	Value of the element	Name of the element	Value of the element	Name of the element	Value of the element
Ck6+CS1L	1.98E-07 F	Lk6+CS1L	8.25E-07 H	Rk6+CS1L	4.57E-04 Ω
Ck6+CS1U	2.09E-07 F	Lk6+CS1U	8.70E-07 H	Rk6+CS1U	4.83E-04 Ω
Ck6+CS2L	1.01E-07 F	Lk6+CS2L	4.20E-07 H	Rk6+CS2L	2.33E-04 Ω
Ck6+CS2U	1.58E-07 F	Lk6+CS2U	6.60E-07 H	Rk6+CS2U	3.66E-04 Ω
Ck6+CS3L	7.20E-08 F	Lk6+CS3L	3.00E-07 H	Rk6+CS3L	1.67E-04 Ω
Ck6+CS3U	1.30E-07 F	Lk6+CS3U	5.40E-07 H	Rk6+CS3U	3.00E-04 Ω
Ck6+PF1	3.60E-08 F	Lk6+PF1	1.50E-07 H	Rk6+PF1	8.38E-05 Ω
Ck6+PF6	3.60E-08 F	Lk6+PF6	1.50E-07 H	Rk6+PF6	8.38E-05 Ω
Ck6-CS1L	1.80E-07 F	Lk6-CS1L	7.50E-07 H	Rk6-CS1L	4.17E-04 Ω
Ck6-CS1U	2.09E-07 F	Lk6-CS1U	8.70E-07 H	Rk6-CS1U	4.83E-04 Ω
Ck6-CS2L	1.01E-07 F	Lk6-CS2L	4.20E-07 H	Rk6-CS2L	2.33E-04 Ω
Ck6-CS2U	1.58E-07 F	Lk6-CS2U	6.60E-07 H	Rk6-CS2U	3.66E-04 Ω
Ck6-CS3L	7.20E-08 F	Lk6-CS3L	3.00E-07 H	Rk6-CS3L	1.67E-04 Ω
Ck6-CS3U	1.30E-07 F	Lk6-CS3U	5.40E-07 H	Rk6-CS3U	3.00E-04 Ω
Ck6-PF1	3.60E-08 F	Lk6-PF1	1.50E-07 H	Rk6-PF1	8.33E-05 Ω
Ck6-PF6	3.60E-08 F	Lk6-PF6	1.50E-07 H	Rk6-PF6	8.33E-05 Ω

Tab. A.9: Capacitances, inductances and resistances used in the network model of the CS PF coil system for the connection of fast discharge units (FDU) with fast discharge resistors by two parallel power cables.

Name of the element	Value of the element	Name of the element	Value of the element	Name of the element	Value of the element
Ck2+CS1L	6.00E-08 F	Lk2+CS1L	2.25E-06 H	Rk2+CS1L	1.25E-03 Ω
Ck2+CS1U	2.05E-07 F	Lk2+CS1U	8.50E-07 H	Rk2+CS1U	4.75E-04 Ω
Ck2+CS2L	4.00E-08 F	Lk2+CS2L	1.48E-06 H	Rk2+CS2L	8.25E-04 Ω
Ck2+CS2U	5.64E-08 F	Lk2+CS2U	2.11E-06 H	Rk2+CS2U	1.18E-03 Ω
Ck2+CS3L	3.60E-08 F	Lk2+CS3L	1.35E-06 H	Rk2+CS3L	7.50E-04 Ω
Ck2+CS3U	4.68E-08 F	Lk2+CS3U	1.75E-06 H	Rk2+CS3U	9.75E-04 Ω
Ck2+PF1	2.40E-08 F	Lk2+PF1	9.00E-07 H	Rk2+PF1	5.00E-04 Ω
Ck2+PF2	3.60E-08 F	Lk2+PF2	1.35E-06 H	Rk2+PF2	7.50E-04 Ω
Ck2+PF3	1.05E-08 F	Lk2+PF3	1.57E-06 H	Rk2+PF3	8.75E-04 Ω
Ck2+PF4	6.00E-09 F	Lk2+PF4	9.00E-07 H	Rk2+PF4	5.00E-04 Ω
Ck2+PF5	6.00E-09 F	Lk2+PF5	9.00E-07 H	Rk2+PF5	5.00E-04 Ω
Ck2+PF6	6.00E-08 F	Lk2+PF6	1.13E-06 H	Rk2+PF6	6.25E-04 Ω
Ck2-CS1L	9.60E-08 F	Lk2-CS1L	3.60E-06 H	Rk2-CS1L	2.00E-03 Ω
Ck2-CS1U	2.12E-07 F	Lk2-CS1U	8.85E-07 H	Rk2-CS1U	4.91E-04 Ω
Ck2-CS2L	4.00E-08 F	Lk2-CS2L	1.48E-06 H	Rk2-CS2L	8.25E-04 Ω
Ck2-CS2U	5.64E-08 F	Lk2-CS2U	2.11E-06 H	Rk2-CS2U	1.18E-03 Ω
Ck2-CS3L	3.80E-08 F	Lk2-CS3L	1.44E-06 H	Rk2-CS3L	8.00E-04 Ω
Ck2-CS3U	4.68E-08 F	Lk2-CS3U	1.75E-06 H	Rk2-CS3U	9.75E-04 Ω
Ck2-PF1	2.40E-08 F	Lk2-PF1	9.00E-07 H	Rk2-PF1	5.00E-04 Ω
Ck2-PF2	3.60E-08 F	Lk2-PF2	1.35E-06 H	Rk2-PF2	7.50E-04 Ω
Ck2-PF3	1.05E-08 F	Lk2-PF3	1.57E-06 H	Rk2-PF3	8.75E-04 Ω
Ck2-PF4	1.26E-08 F	Lk2-PF4	1.89E-06 H	Rk2-PF4	1.00E-03 Ω
Ck2-PF5	1.30E-08 F	Lk2-PF5	1.89E-06 H	Rk2-PF5	1.00E-06 Ω
Ck2-CS1L	9.60E-08 F	Lk2-PF6	2.11E-06 H	Rk2-PF6	1.18E-03 Ω

Tab. A.10:Capacitance, inductances and resistances used in the network model of the CS PF coil system for the connection of the coil terminal box (CTB) with coil terminals by super-conducting bus bars

Name of the element	Value of the element	Name of the element	Value of the element	Name of the element	Value of the element
Cscbb+CS1L	7.20E-08 F	Lscbb+CS1L	5.60E-05 H	Rscbb+CS1L	4.00E-11 Ω
Cscbb+CS1U	7.20E-08 F	Lscbb+CS1U	5.60E-05 H	Rscbb+CS1U	4.00E-11 Ω
Cscbb+CS2L	6.98E-08 F	Lscbb+CS2L	5.46E-05 H	Rscbb+CS2L	3.90E-11 Ω
Cscbb+CS2U	6.40E-08 F	Lscbb+CS2U	5.04E-05 H	Rscbb+CS2U	3.60E-11 Ω
Cscbb+CS3L	6.60E-08 F	Lscbb+CS3L	5.18E-05 H	Rscbb+CS3L	3.70E-11 Ω
Cscbb+CS3U	6.10E-08 F	Lscbb+CS3U	4.76E-05 H	Rscbb+CS3U	3.40E-11 Ω
Cscbb+PF1	4.48E-08 F	Lscbb+PF1	3.50E-05 H	Rscbb+PF1	2.50E-11 Ω
Cscbb+PF2	4.50E-08 F	Lscbb+PF2	3.50E-05 H	Rscbb+PF2	2.50E-11 Ω
Cscbb+PF3	4.10E-08 F	Lscbb+PF3	3.20E-05 H	Rscbb+PF3	2.30E-11 Ω
Cscbb+PF4	4.80E-08 F	Lscbb+PF4	3.80E-05 H	Rscbb+PF4	2.70E-11 Ω
Cscbb+PF5	4.60E-08 F	Lscbb+PF5	3.60E-05 H	Rscbb+PF5	2.60E-11 Ω
Cscbb+PF6	4.48E-08 F	Lscbb+PF6	3.50E-05 H	Rscbb+PF6	2.50E-11 Ω
Cscbb-CS1L	7.20E-08 F	Lscbb-CS1L	5.60E-05 H	Rscbb-CS1L	4.00E-11 Ω
Cscbb-CS1U	7.20E-08 F	Lscbb-CS1U	5.60E-05 H	Rscbb-CS1U	4.00E-11 Ω
Cscbb-CS2L	6.98E-08 F	Lscbb-CS2L	5.46E-05 H	Rscbb-CS2L	3.90E-11 Ω
Cscbb-CS2U	6.40E-08 F	Lscbb-CS2U	5.04E-05 H	Rscbb-CS2U	3.60E-11 Ω
Cscbb-CS3L	6.60E-08 F	Lscbb-CS3L	5.18E-05 H	Rscbb-CS3L	3.70E-11 Ω
Cscbb-CS3U	6.10E-08 F	Lscbb-CS3U	4.76E-05 H	Rscbb-CS3U	3.40E-11 Ω
Cscbb-PF1	4.48E-08 F	Lscbb-PF1	3.50E-05 H	Rscbb-PF1	2.50E-11 Ω
Cscbb-PF2	4.50E-08 F	Lscbb-PF2	3.50E-05 H	Rscbb-PF2	2.50E-11 Ω
Cscbb-PF3	4.10E-08 F	Lscbb-PF3	3.20E-05 H	Rscbb-PF3	2.30E-11 Ω
Cscbb-PF4	4.80E-08 F	Lscbb-PF4	3.80E-05 H	Rscbb-PF4	2.70E-11 Ω
Cscbb-PF5	4.60E-08 F	Lscbb-PF5	3.00E-05 H	Rscbb-PF5	2.60E-11 Ω
Cscbb-PF6	4.48E-08 F	Lscbb-PF6	3.50E-05 H	Rscbb-PF6	2.50E-11 Ω

Detailed Network Model of PF 3 Coil

Tab. A.11:Calculated capacitances between the turns of PF 3 coil.

Capacitances of turns between the layers in one double pancake		Capacitances of turns between double pancakes		Capacitances between the turns of one layer	
C_P01_02_C01	1.83E-08 F	C_P02_03_C01	1.44E-08 F	C_P01_C01_02	1.97E-08 F
C_P01_02_C02	1.84E-08 F	C_P02_03_C02	1.45E-08 F	C_P01_C02_03	1.98E-08 F
C_P01_02_C03	1.85E-08 F	C_P02_03_C03	1.45E-08 F	C_P01_C03_04	1.99E-08 F
C_P01_02_C04	1.86E-08 F	C_P02_03_C04	1.46E-08 F	C_P01_C04_05	2.00E-08 F
C_P01_02_C05	1.87E-08 F	C_P02_03_C05	1.47E-08 F	C_P01_C05_06	2.01E-08 F
C_P01_02_C06	1.88E-08 F	C_P02_03_C06	1.48E-08 F	C_P01_C06_07	2.02E-08 F
C_P01_02_C07	1.88E-08 F	C_P02_03_C07	1.48E-08 F	C_P01_C07_08	2.03E-08 F
C_P01_02_C08	1.89E-08 F	C_P02_03_C08	1.49E-08 F	C_P01_C08_09	2.04E-08 F
C_P01_02_C09	1.90E-08 F	C_P02_03_C09	1.50E-08 F	C_P01_C09_10	2.05E-08 F
C_P01_02_C10	1.91E-08 F	C_P02_03_C10	1.51E-08 F	C_P01_C10_11	2.06E-08 F
C_P01_02_C11	1.92E-08 F	C_P02_03_C11	1.51E-08 F	C_P01_C11_12	2.07E-08 F
C_P01_02_C12	1.93E-08 F	C_P02_03_C12	1.52E-08 F		

Tab. A.12:Calculated capacitances of turns to ground of PF 3 coil.

Capacitances of the turns to ground in lowest and uppermost pancakes		Capacitance to ground for outer turns on the left and right side of the coil		Capacitance to ground for lowest and uppermost outer turns in Z-direction	
C_P01_E_C01	1.98E-08 F	C_P02_E_C01	1.02E-08 F	C_P01_E_C01	9.53E-09 F
C_P01_E_C02	9.58E-09 F	C_P02_E_C12	1.05E-08 F	C_P01_E_C12	1.01E-08 F
C_P01_E_C03	9.63E-09 F				
C_P01_E_C04	9.68E-09 F				
C_P01_E_C05	9.72E-09 F				
C_P01_E_C06	9.77E-09 F				
C_P01_E_C07	9.82E-09 F				
C_P01_E_C08	9.87E-09 F				
C_P01_E_C09	9.92E-09 F				
C_P01_E_C10	9.96E-09 F				
C_P01_E_C11	1.00E-08 F				
C_P01_E_C12	2.05E-08 F				

Detailed Network Model of PF 6 Coil

Tab. A.13:Calculated capacitances between the turns of PF 6 coil.

Capacitances of turns between the layers in one double pancake		Capacitances of turns between double pancakes		Capacitances between the turns of one layer	
C_P01_02_C01	5.84E-09 F	C_P02_03_C01	5.12E-09 F	C_P01_C01_02	6.06E-09 F
C_P01_02_C02	5.94E-09 F	C_P02_03_C02	5.21E-09 F	C_P01_C02_03	6.17E-09 F
C_P01_02_C03	6.04E-09 F	C_P02_03_C03	5.30E-09 F	C_P01_C03_04	6.27E-09 F
C_P01_02_C04	6.14E-09 F	C_P02_03_C04	5.39E-09 F	C_P01_C04_05	6.38E-09 F
C_P01_02_C05	6.25E-09 F	C_P02_03_C05	5.48E-09 F	C_P01_C05_06	6.48E-09 F
C_P01_02_C06	6.35E-09 F	C_P02_03_C06	5.56E-09 F	C_P01_C06_07	6.58E-09 F
C_P01_02_C07	6.45E-09 F	C_P02_03_C07	5.65E-09 F	C_P01_C07_08	6.69E-09 F
C_P01_02_C08	6.55E-09 F	C_P02_03_C08	5.74E-09 F	C_P01_C08_09	6.79E-09 F
C_P01_02_C09	6.65E-09 F	C_P02_03_C09	5.83E-09 F	C_P01_C09_10	6.90E-09 F
C_P01_02_C10	6.75E-09 F	C_P02_03_C10	5.92E-09 F	C_P01_C10_11	7.00E-09 F
C_P01_02_C11	6.86E-09 F	C_P02_03_C11	6.01E-09 F	C_P01_C11_12	7.11E-09 F
C_P01_02_C12	6.96E-09 F	C_P02_03_C12	6.10E-09 F	C_P01_C12_13	7.21E-09 F
C_P01_02_C13	7.06E-09 F	C_P02_03_C13	6.19E-09 F	C_P01_C13_14	7.32E-09 F
C_P01_02_C14	7.16E-09 F	C_P02_03_C14	6.28E-09 F	C_P01_C14_15	7.42E-09 F
C_P01_02_C15	7.26E-09 F	C_P02_03_C15	6.37E-09 F	C_P01_C15_16	7.53E-09 F
C_P01_02_C16	7.36E-09 F	C_P02_03_C16	6.46E-09 F	C_P01_C16_17	7.63E-09 F
C_P01_02_C17	7.47E-09 F	C_P02_03_C17	6.54E-09 F	C_P01_C17_18	7.73E-09 F
C_P01_02_C18	7.57E-09 F	C_P02_03_C18	6.63E-09 F	C_P01_C18_19	7.84E-09 F
C_P01_02_C19	7.67E-09 F	C_P02_03_C19	6.72E-09 F	C_P01_C19_20	7.94E-09 F
C_P01_02_C20	7.77E-09 F	C_P02_03_C20	6.81E-09 F	C_P01_C20_21	8.05E-09 F
C_P01_02_C21	7.87E-09 F	C_P02_03_C21	6.90E-09 F	C_P01_C21_22	8.15E-09 F
C_P01_02_C22	7.97E-09 F	C_P02_03_C22	6.99E-09 F	C_P01_C22_23	8.26E-09 F
C_P01_02_C23	8.08E-09 F	C_P02_03_C23	7.08E-09 F	C_P01_C23_24	8.36E-09 F
C_P01_02_C24	8.18E-09 F	C_P02_03_C24	7.17E-09 F	C_P01_C24_25	8.47E-09 F
C_P01_02_C25	8.28E-09 F	C_P02_03_C25	7.26E-09 F	C_P01_C25_26	8.57E-09 F
C_P01_02_C26	8.38E-09 F	C_P02_03_C26	7.35E-09 F	C_P01_C26_27	8.68E-09 F
C_P01_02_C27	8.48E-09 F	C_P02_03_C27	7.44E-09 F		

Tab. A.14:Calculated capacitances of turns to ground of PF 6 coil.

Capacitances of the turns to ground in lowest and uppermost pancakes		Capacitance to ground for outer turns on the left and right side of the coil		Capacitance to ground for lowest and uppermost outer turns in Z-direction	
C_P01_E_C01	6.25E-09 F	C_P02_E_C01	3.11E-09 F	C_P01_E_C01	3.14E-09 F
C_P01_E_C02	3.20E-09 F	C_P02_E_C27	4.43E-09 F	C_P01_E_C27	4.56E-09 F
C_P01_E_C03	3.25E-09 F				
C_P01_E_C04	3.31E-09 F				
C_P01_E_C05	3.36E-09 F				
C_P01_E_C06	3.41E-09 F				
C_P01_E_C07	3.47E-09 F				
C_P01_E_C08	3.52E-09 F				
C_P01_E_C09	3.58E-09 F				
C_P01_E_C10	3.63E-09 F				
C_P01_E_C11	3.69E-09 F				
C_P01_E_C12	3.74E-09 F				
C_P01_E_C13	3.80E-09 F				
C_P01_E_C14	3.85E-09 F				
C_P01_E_C15	3.91E-09 F				
C_P01_E_C16	3.96E-09 F				
C_P01_E_C17	4.02E-09 F				
C_P01_E_C18	4.07E-09 F				
C_P01_E_C19	4.13E-09 F				
C_P01_E_C20	4.18E-09 F				
C_P01_E_C21	4.23E-09 F				
C_P01_E_C22	4.29E-09 F				
C_P01_E_C23	4.34E-09 F				
C_P01_E_C24	4.40E-09 F				
C_P01_E_C25	4.45E-09 F				
C_P01_E_C26	4.51E-09 F				
C_P01_E_C27	8.99E-09 F				

A.4   Voltage Waveforms Calculated within PF 3 Coil

Fast Discharge

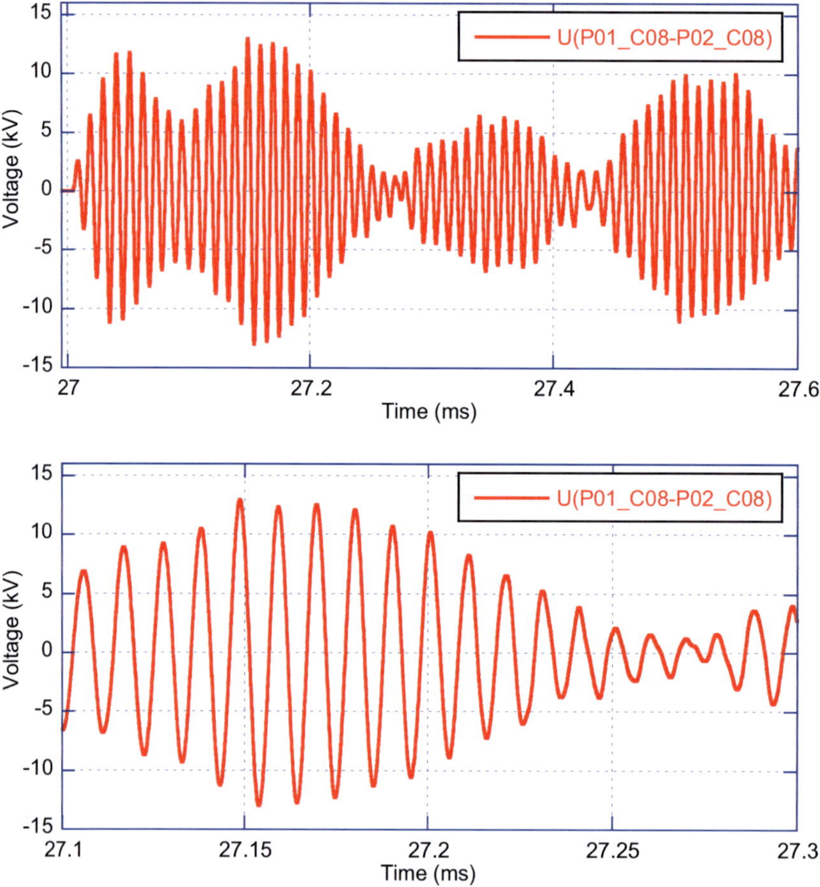

Fig. A.6:   Voltage waveforms of maximum calculated voltage on layer insulation between turns P01_C08 and P02_C08 during fast discharge. Calculated with 0.51 kHz detailed network model of PF 3 coil. Maximum time of sampling steps 0.1 μs. Upper part in time scale between 27 ms and 27.6 ms, lower part in time scale between 27.1 ms and 27.3 ms.

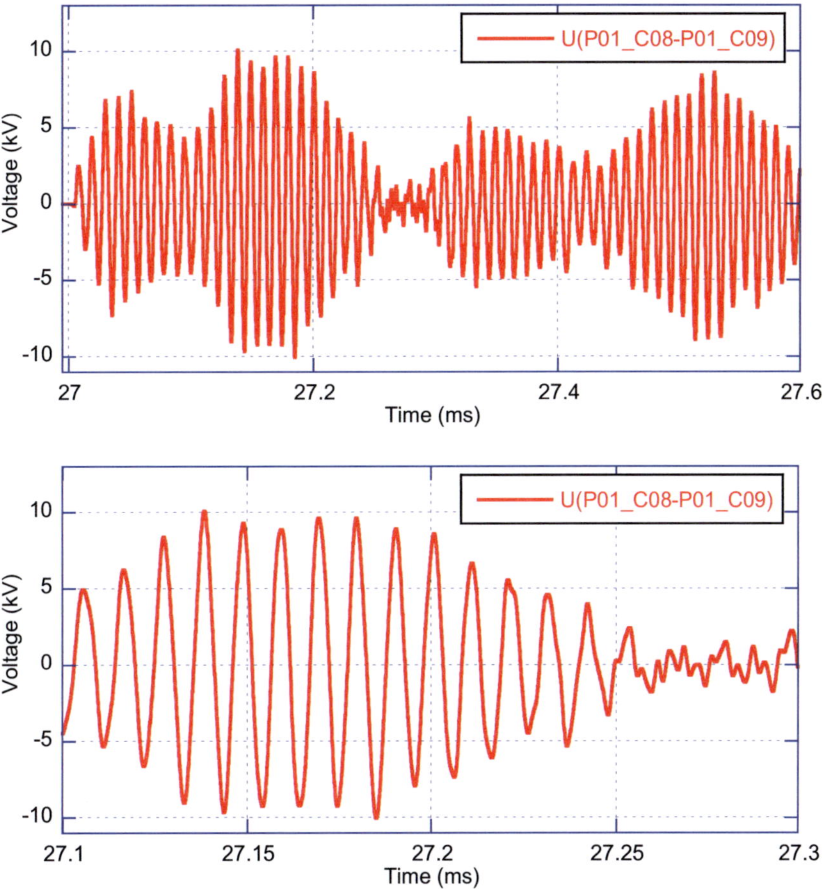

Fig. A.7:   Voltage waveforms of maximum calculated voltage on turn insulation between turns P01_C08 and P01_C09 during fast discharge. Calculated with 0.51 kHz detailed network model of PF 3 coil. Maximum time of sampling steps 0.1 μs. Upper part in time scale between 27 ms and 27.6 ms, lower part in time scale between 27.1 ms and 27.3 ms.

Failure Case 1

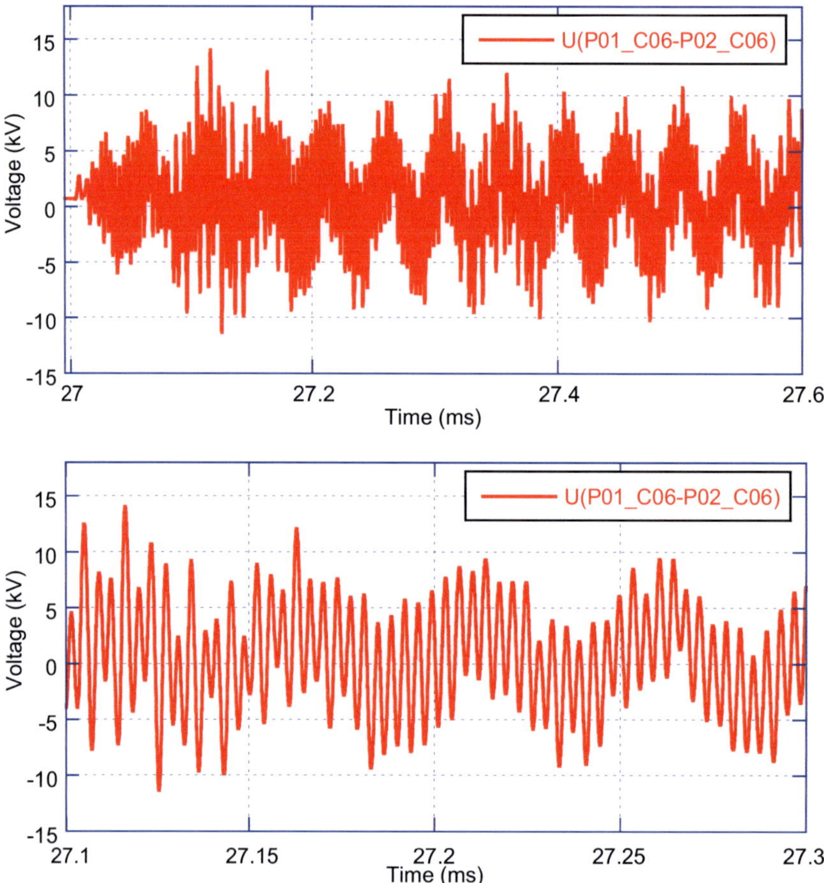

Fig. A.8:   Voltage waveforms of maximum calculated voltage on layer insulation between turns P01_C06 and P02_C06 during fast discharge. Calculated with 25.5 kHz detailed network model of PF 3 coil. Maximum time of sampling steps 0.1 µs. Upper part in time scale between 27 ms and 27.6 ms, lower part in time scale between 27.1 ms and 27.3 ms.

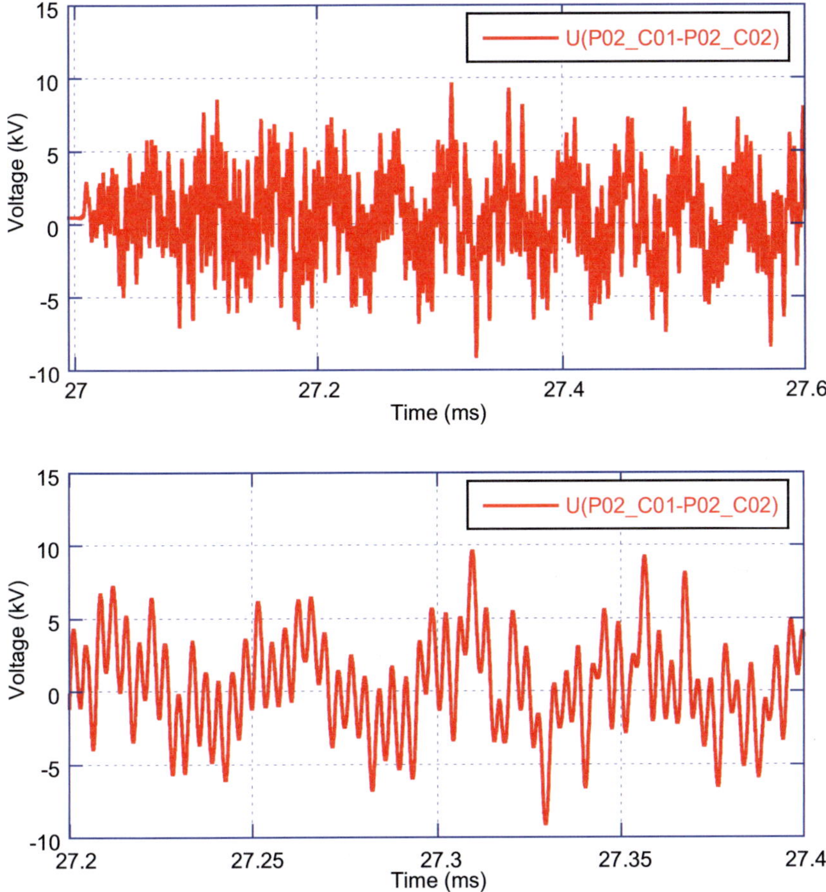

Fig. A.9: Voltage waveforms of maximum calculated voltage turn insulation between turns P02_C01 and P02_C02 during fast discharge. Calculated with 25.5 kHz detailed network model of PF 3 coil. Maximum time of sampling steps 0.1 µs. Upper part in time scale between 27 ms and 27.6 ms, lower part in time scale between 27.2 ms and 27.4 ms.

Failure Case 2

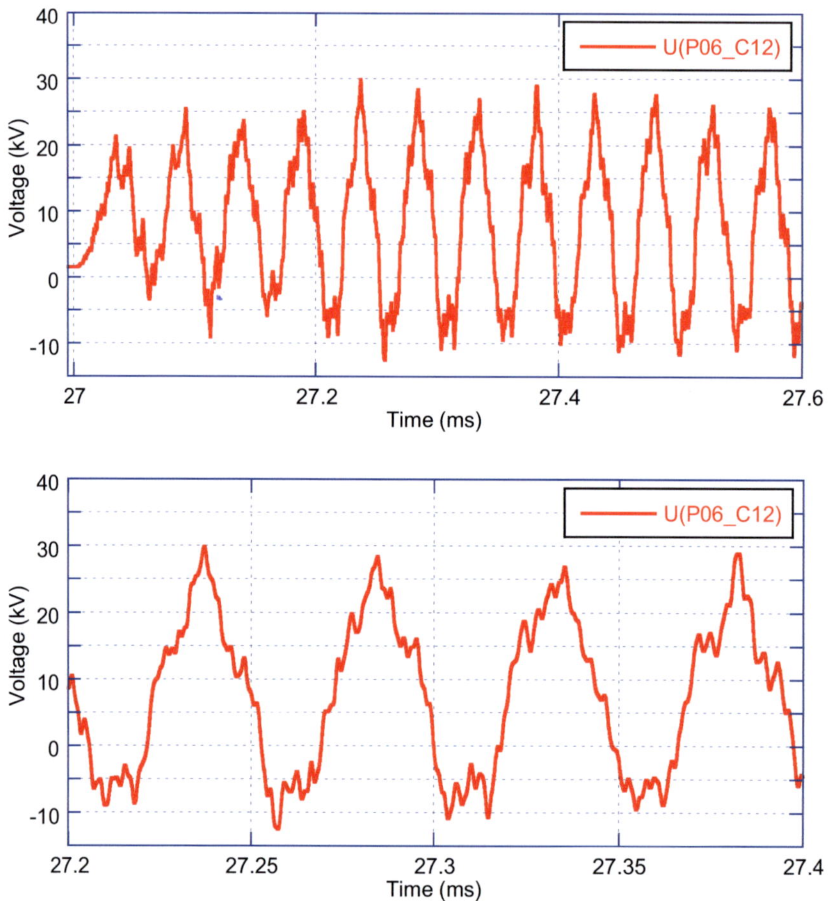

Fig. A.10: Voltage waveforms of maximum calculated voltage on layer insulation between turns P14_C01 and P15_C01 during fast discharge. Calculated with 25.5 kHz detailed network model of PF 3 coil. Maximum time of sampling steps 0.1 µs. Upper part in time scale between 27 ms and 27.6 ms, lower part in time scale between 27.1 ms and 27.3 ms.

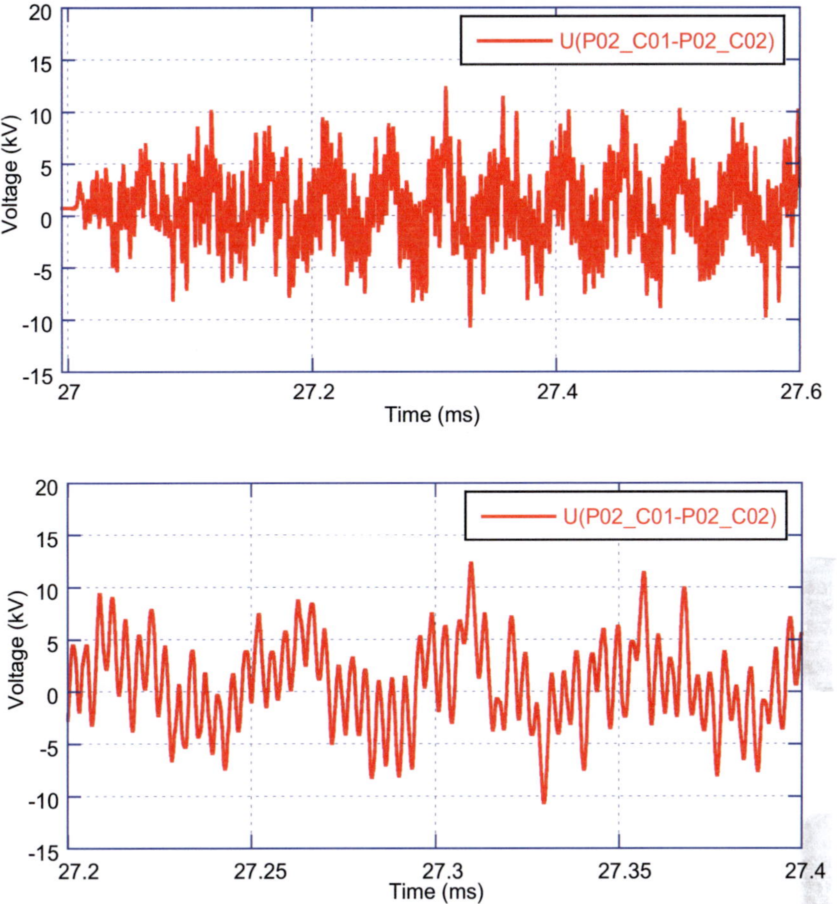

Fig. A.11: Voltage waveforms of maximum calculated voltage on turn insulation between turns P02_C01 and P02_C02 at failure case 2. Calculated with 25.5 kHz detailed network model of PF 3 coil. Maximum time of sampling steps 0.1 µs. Upper part in time scale between 27 ms and 27.6 ms, lower part in time scale between 27.2 ms and 27.4 ms.

Tab. A.15:Calculated maximum voltages with detailed network model of PF_3 coil in the reference scenario and during fast discharge.

		0.51 kHz				25.5 kHz				200 kHz			
		Terminal to ground	Ground insulation	Layer insulation	Turn insulation	Terminal to ground	Ground insulation	Layer insulation	Turn insulation	Terminal to ground	Ground insulation	Layer insulation	Turn insulation
Reference scenario	Maximum voltage	-14.9 kV	-14.9 kV	-2.6 kV	0.9 kV	-14.9 kV	-14.9 kV	-2.6 kV	0.9 kV	-14.9 kV	-14.9 kV	-2.6 kV	0.9 kV
	Location	Positive terminal	P01_C12 to ground	P12_C01 to P13_C01	P01_C01 to P01_C02	Positive terminal	P01_C12 to ground	P12_C01 to P13_C01	P01_C01 to P01_C02	Positive terminal	P01_C12 to ground	P12_C01 to P13_C01	P01_C01 to P01_C02
	Time	607 ms	607 ms	608 ms	608 ms	607 ms	607 ms	608 ms	608 ms	607 ms	607 ms	608 ms	608 ms
	Rise time	2.2 ms	2.2 ms	2.2 ms	2.2 ms	2.2 ms	2.2 ms	2.2 ms	2.2 ms	2.2 ms	2.2 ms	2.2 ms	2.2 ms
	Start rise voltage	-13.8 kV	-13.8 kV	-2.4 kV	0.8 kV	-13.8 kV	-13.8 kV	-2.4 kV	0.8 kV	-13.8 kV	-13.8 kV	-2.4 kV	0.8 kV
		Terminal to ground	Ground insulation	Layer insulation	Turn insulation	Terminal to ground	Ground insulation	Layer insulation	Turn insulation	Terminal to ground	Ground insulation	Layer insulation	Turn insulation
Fast discharge	Maximum voltage	-5.3 kV	-12.1 kV	-13 kV	10 kV	-5.3 kV	7.0 kV	8.2 kV	-4.9 kV	-5.3 kV	12.3 kV	-11.3 kV	-8.3 kV
	Location	Negative terminal	P01_C06 to ground	P01_C08 to P02_C08	P01_C08 to P01_C09	Negative terminal	P01_C02 to ground	P08_C02 to P09_C02	P14_C10 to P14_C11	Negative terminal	P13_C12	P08_C02 to P09_C02	P04_C11 to P04_C12
	Time	27.007 ms	27.523 ms	27.154 ms	27.138 ms	27.007 ms	27.105 ms	27.068 ms	27.104 ms	27.007 ms	27.301 ms	27.290 ms	27.251 ms
	Rise time	3.5 μs	2.97 μs	3.05 μs	3.4 μs μs	3.5 μs	3.5 μs	3.8 μs	3.4 μs	3.5 μs	3.3 μs	3.3 μs	3.5 μs
	Start rise voltage	0.0 kV	8.7 kV	12.9 kV	-9.0 kV	0.0 kV	-3.7 kV	-7.1 kV	3.6 kV	0.0 kV	-10.2 kV	11.2 kV	7.0 kV

Tab. A.16:Calculated maximum voltages with detailed network model of PF 3 coil at failure case 1 and failure case 2.

		0.51 kHz				25.5 kHz				200 kHz			
		Terminal to ground	Ground insulation	Layer insulation	Turn insulation	Terminal to ground	Ground insulation	Layer insulation	Turn insulation	Terminal to ground	Ground insulation	Layer insulation	Turn insulation
Failure case 1	Maximum voltage	19.3 kV	23.4 kV	10.8 kV	7.6 kV	19.3 kV	21.3 kV	14.0 kV	9.6 kV	19.3 kV	20.8 kV	8.4 kV	-8.4 kV
	Location	Positive terminal	P01_C06 to ground	P02_C03 to P03_C03	P02_C09 to P02_C10	Positive terminal	P02_C01 to ground	P01_C06 to P02_C06	P02_C01 to P02_C02	Positive terminal	P01_C06 to ground	P15_C08 to P16_C08	P02_C04 to P02_C05
	Time	27.031 ms	27.031 ms	27.080 ms	27.074 ms	27.031 ms	27.031 ms	27.116 ms	27.310 ms	27.031 ms	27.086 ms	27.029 ms	27.110 ms
	Rise time	15.5 μs	4.1 μs	15.6 kV	14.9 μs	15.5 μs	15.1 μs	1.1 μs	1.2 μs	15.5 μs	5.1 μs	2.1 μs	2.3 μs
	Start rise voltage	5.1 kV	8.0 kV	-6.5 kV	-5.7 kV	5.1 kV	5.0 kV	-7.2 kV	2.6 kV	5.1 kV	9.5 kV	-1.9 kV	2.0 kV
		Terminal to ground	Ground insulation	Layer insulation	Turn insulation	Terminal to ground	Ground insulation	Layer insulation	Turn insulation	Terminal to ground	Ground insulation	Layer insulation	Turn insulation
Failure case 2	Maximum voltage	25.4 kV	29.5 kV	13.9 kV	9.9 kV	25.4 kV	29.8 kV	17.1 kV	12.3 kV	25.4 kV	27.5 kV	11.4 kV	-10.5 kV
	Location	Positive terminal	P01_C06 to ground	P01_C08 to P02_C08	P01_C06 to P01_C07	Positive terminal	P06_C12 to ground	P14_C01 to P15_C01	P02_C01 to P02_C02	Positive terminal	P01_C06 to ground	P15_C06 to P16_C06	P02_C04 to P02_C05
	Time	27.031 ms	27.031 ms	27.177 ms	27.188 ms	27.031 ms	27.237 ms	27.380 ms	27.310 ms	27.031 ms	27.086 ms	27.751 ms	27.110 ms
	Rise time	15.8 μs	4.2 μs	3.6 μs	4.4 μs	15.8 μs	16.5 μs	13.4 μs	4.3 μs	15.8 μs	5.3 μs	1.6 μs	1.5 μs
	Start rise voltage	7.0 kV	13.1 kV	-3.8 kV	-2.5 kV	7.0 kV	-8.7 kV	-8.7 kV	-3.0 kV	7.0 kV	14.4 kV	-1.2 kV	0.7 kV

Tab. A.17:Calculated maximum voltages with detailed network model of PF 6 coil in the reference scenario and during fast discharge.

		0.55 kHz				32 kHz				200 kHz			
		Terminal to ground	Ground insulation	Layer insulation	Turn insulation	Terminal to ground	Ground insulation	Layer insulation	Turn insulation	Terminal to ground	Ground insulation	Layer insulation	Turn insulation
Reference scenario	Maximum voltage	7.2 kV	7.2 kV	-2.6 kV	-0.9 kV	7.2 kV	7.2 kV	-2.6 kV	-0.9 kV	7.2 kV	7.2 kV	-2.6 kV	-0.9 kV
	Location	Negative terminal	P16_C27 to ground	P14_C02 to P15_C02	P04_C26 to P04_C27	Negative terminal	P16_C27 to ground	P14_C02 to P15_C02	P04_C26 to P04_C27	Negative terminal	P16_C27 to ground	P14_C02 to P15_C02	P04_C26 to P04_C27
	Time	20 ms	20 ms	20 ms	20 ms	20 ms	20 ms	20 ms	20 ms	20 ms	20 ms	20 ms	20 ms
	Rise time	4ms	4 ms	4 ms	4 ms	4ms	4 ms	4 ms	4 ms	4ms	4 ms	4 ms	4 ms
	Start rise voltage	6.3 kV	6.3 kV	-2.3 kV	-0.8 kV	6.3 kV	6.3 kV	-2.3 kV	-0.8 kV	6.3 kV	6.3 kV	-2.3 kV	-0.8 kV
		Terminal to ground	Ground insulation	Layer insulation	Turn insulation	Terminal to ground	Ground insulation	Layer insulation	Turn insulation	Terminal to ground	Ground insulation	Layer insulation	Turn insulation
Fast discharge	Maximum voltage	-5.0 kV	-8.4 kV	7.6 kV	4.9 kV	-5.0 kV	7.2 kV	5.8 kV	4.9 kV	-5.0 kV	-9.8 kV	-7.4 kV	7.5 kV
	Location	Negative terminal	P16_C19 to ground	P14_C19 to P15_C19	P01_C13 to P01_C14	Negative terminal	P01_C17 to ground	P01_C17 to P02_C17	P16_C22 to P16_C23	Negative terminal	P03_C01 to ground	P02_C09 to P03_C09	P16_C24 to P16_C25
	Time	27.027 ms	27.017 ms	27.279 ms	27.0	27.027 ms	27.015 ms	27.020 ms	27.036 ms	27.027 ms	27.348 ms	27.353 ms	27.077 ms
	Rise time	2.7 µs	2.0 µs	1.7 µs	1.3 µs	2.7 µs	1.0 µs	0.8 µs	1.3 µs	2.7 µs	1.7 µs	1.8 µs	2.4 µs
	Start rise voltage	1.0 kV	3.0 kV	-6.3 kV	-1.5 kV	1.0 kV	-3.4 kV	-5.4 kV	-2.9 kV	1.0 kV	5.6 kV	2.7 kV	-5.0 kV

Tab. A.18:Calculated maximum voltages with detailed network model of PF 6 coil at failure case 1 and failure case 2.

		0.55 kHz				32 kHz				200 kHz			
		Terminal to ground	Ground insulation	Layer insulation	Turn insulation	Terminal to ground	Ground insulation	Layer insulation	Turn insulation	Terminal to ground	Ground insulation	Layer insulation	Turn insulation
Failure case 1	Maximum voltage	8.5 kV	10.1 kV	5.7 kV	4.5 kV	8.5 kV	12.0 kV	5.9 kV	4.5 kV	8.5 kV	11.1 kV	6.2 kV	4.9 kV
	Location	Positive terminal	P01_C13 to ground	P15_C13 to P16_C13	P15_C05 to P15_C06	Positive terminal	P01_C13 to ground	P01_C15 to P02_C15	P14_C19 to P14_C20	Positive terminal	P01_C15 to ground	P01_C19 to P02_C19	P01_C17 to P01_C18
	Time	27.025 ms	27.023 ms	27.485 ms	27.012 ms	27.025 ms	27.024 ms	27.051 ms	27.462 ms	27.025 ms	27.023 ms	27.337 ms	27.849 ms
	Rise time	12.3 µs	10.4 µs	1.8 µs	2.5 µs	12.3 µs	1.4 µs	1.3 µs	1.9 µs	12.3 µs	1.3 µs	1.4 µs	1.2 µs
	Start rise voltage	2.3 kV	1.9 kV	-3.8 kV	-0.4 kV	2.3 kV	3.7 kV	-2.2 kV	-1.5 kV	2.3 kV	4.7 kV	-3.2 kV	-1.8 kV
		Terminal to ground	Ground insulation	Layer insulation	Turn insulation	Terminal to ground	Ground insulation	Layer insulation	Turn insulation	Terminal to ground	Ground insulation	Layer insulation	Turn insulation
Failure case 2	Maximum voltage	23.4 kV	25.9 kV	12.0 kV	8.9 kV	23.4 kV	30.1 kV	13.1 kV	9.0 kV	23.4 kV	27.7 kV	11.6 kV	8.0 kV
	Location	Positive terminal	P01_C01 to ground	P02_C12 to P03_C12	P16_C12 to P16_C13	Positive terminal	P01_C15 to ground	P14_C02 to P15_C02	P16_C18 to P16_C19	Positive terminal	P01_C15 to ground	P14_C02 to P15_C02	P16_C16 to P16_C17
	Time	27.025 ms	27.029 ms	27.058 ms	27.445 ms	27.025 ms	27.024 ms	27.207 ms	27.245 ms	27.025 ms	27.023 ms	27.135 ms	27.287 ms
	Rise time	10.4 µs	15.6 µs	11.9 µs	2.3 µs	10.4 µs	1.5 µs	17.1 µs	1.9 µs	10.4 µs	1.4 µs	17.4 µs	2.1 µs
	Start rise voltage	7.2 kV	7.9 kV	-6.7 kV	-1.5 kV	7.2 kV	14.5 kV	-5.4 kV	-1.1 kV	7.2 kV	15.8 kV	-5.0 kV	0.1 kV

B Designations and Abbreviations

AC	Alternating Current, (A)
BC	Booster Converter
BPS	Bypass Switch
C_C	Counterpulse Capacitor, (F)
C_{cyl}	Capacitance of Cylinder Capacitor, (F)
CCU	Current Commutation Unit
C_{pp}	Capacitance of Parallel Plate Capacitor, (F)
CS coil	Central Solenoid Coil
CTB	Coil Terminal Box
Cu	Copper
D	Diode
d	Distance, (m)
DC	Direct Current, (A)
EPMS	Explosively Activated Protective Make Switch
f	Frequency, (Hz)
F24	24 Filament coil
F54	54 Filament coil
FDU	Fast Discharge Unit
FEM	Finite Element Method
FFT	Fast Fourier Transformation
FG	Frequency Generator
h_c	Height of the Capacitor, (m)
IC	Instrumentation Cable
ITEP	Institute of Technical Physics
ITER	International Thermonuclear Experimental Reactor
KIT	Karlsruhe Institute of Technology
L_C	Inductance of Coil, (H)
L_{CS}	Inductance of Cross Section, (H)
L_S	Saturable Inductance, (H)
MC	Main Converter
$M_{C1\text{-}C2}$	Mutual Inductance between Coils, (H)
$M_{CS1\text{-}CS2}$	Mutual Inductance between Cross Sections of Coils, (H)
MS	Main Switch
NbTi	Niobium-Titan
Nb_3Sn	Niobium-3-Tinn
PB	Pyrobreaker
PF coil	Poloidal Field coil
PMS	Protective Make Switch
PS	Power Supply
R_D	Discharge Resistor, (Ω)
r_i	Inner Radius, (m)
R_{NG}	Resistance Neutral to Ground, (Ω)
r_o	Outer Radius, (m)
RT	Room Temperature
R_{TN}	Resistance Terminal to Neutral, (Ω)
S	Switch
SC	Snubber Circuit
SNU	Switching Network Unit
SNR	Switching Network Resistor, (Ω)
TCB	Thyristor Circuit Breaker
TF coil	Toroidal Field Coil
TFMC	Toroidal Field Model Coil
Th	Thyristor
U	Voltage, (V)
VCB	Vacuum Circuit Breaker
VS	Vacuum Switch

VSC	Vertical Stabilisation Converter
X_C	Capacitive reactance, (Ω)
X_L	Inductive reactance, (Ω)
w	Number of Turns

C References

[Bar05] Bareyt, B.: *Electrical Connections from the CCUs to the Resistors and to the Capacitors, (Part of WBS 4.1)*, June 2005, ITER_D_22J2TY

[Bau98] Bauer, K.; Fink S.; Friesinger, G.; Ulbricht, A.; Wüchner, F.: *The electrical insulation system of forced flow cooled superconducting magnet*, Cryogenics 39 (1998), pp. 1123 – 1134

[CDA1] *Design Description Document, DDD 1.1 Magnet, 1.2 Component Description, Annex 1: Feeder Design*

[CDA3] *Design Descirption Document, DDD 1.1 Magnet, 1.2 Component Description, Annex 3: Control and Instrumentation*

[CDA4] *Design Description Document, DDD1.1 Magnet, 1.2 Component Description, Annex 4: Electrical Insulation Design and Monitoring*

[CDA4a] *Design Description Document, DDD1.1 Magnet, 1.2 Component Description, Annex 4: Electrical Insulation Design and Monitoring*, Fig. 3-4 p. 16

[DDD01] *Design Description Document, DDD 1.1 Magnet, Section 1. Engineering Description, 1.1 System Description, 1.2 Componenent Descriptions, 1.3 Magnet Procurment Packages*, N 11 DDD 142 01-08-12 R 0.1

[DDD06] *Design Description Document, DDD 1.1 Magnet, Section 1. Engineering Description*, October 2006

[DDD06a] *Design Description Document, DDD 1.1 Magnet, Section 1. Engineering Description*, October 2006, Tab. 1.2.3-5, p. 103, Tab. 1.2.4-1, p. 110

[DDD06b] *Design Description Document, DDD 1.1 Magnet, Section 1. Engineering Description*, October 2006, Tab. 1.2.4-7, p. 114

[DDD06c] *Design Description Document, DDD 1.1 Magnet, Section 1. Engineering Description*, October 2006, p. 113

[DDD06d] *Design Description Document, DDD 1.1 Magnet, Section 1. Engineering Description*, October 2006, p. 115

[DDD06e]   *Design Description Document, DDD 1.1 Magnet, Section 1. Engineering Description*, October 2006, Tab. 1.2.4-8, p. 116

[DDD06f]   *Design Description Document, DDD 1.1 Magnet, Section 1. Engineering Description*, October 2006, p. 102

[DDD06g]   *Design Description Document, DDD 1.1 Magnet, Section 1. Engineering Description*, October 2006, p. 18

[DDD06h]   *Design Description Document, DDD 1.1 Magnet, Section 1. Engineering Description*, October 2006, Tab. 1.2.6-4, p. 123

[DDD06i]   *Design Description Document, DDD 1.1 Magnet, Section 1. Engineering Description*, October 2006, pp. 120 - 125

[DDD06j]   *Design Description Document, DDD 1.1 Magnet, Section 1. Engineering Description*, October 2006, p. 48

[DDD06k]   *Design Description Document, DDD 1.1 Magnet, Section 1. Engineering Description*, October 2006, p. 9

[DDD06m]   *Design Description Document, DDD 1.1 Magnet, Section 1. Engineering Description*, October 2006, p. 11

[DDDD1]   *Design Description Documents Drawings*, 12.0187.0001

[DDDD2]   *Design Description Documents Drawings*, 12.000438.02, 12.000438.03, 12.000438.04

[DDDD3]   *Design Description Documents Drawings*, 12.000439.01, 12.000439.03, 12.000442.02, 12.000442.06, 12.000380.02, 12.000446.02, 12.000446.05, 13.000373.04, 13.000373.06

[DDDD4]   *Design Description Documents Drawings*, 12.000426.02, 12.000439.02, 12.000439.03, 12.000446.01, 12.000446.02, 12.000446.05, 12.000427.01

[DRG1a]   *Design Requirements and Guidelines Level 1, Magnet Superconducting and Electrical Design Criteria*, N 11 FDR 12 01-07-02 R 0.1, p. 47

[DRGAa]   *Annex to Design Requirements and Guidelines Level 1, Superconducting Material Database, Article 5. Terminal, Electrical and Mechanical Properties of Materials at Cryogenic Temperatures*, p. 3, p. 8

[DRGAb] *Annex to Design Requirements and Guidelines Level 1, Superconducting Material Database, Article 5. Terminal, Electrical and Mechanical Properties of Materials at Cryogenic Temperatures*, p. 37

[Fas70] Fastowski, W. G.; Petrowski, J. W.; Rowinski, A. E.: *Kryotechnik*, Akademie Verlag, Berlin 1970, p. 342

[Feh77] Feher, A.; Reiffers, M.; Petrovic P.; Janos S.: *The Electrical Resistivity of Nb-Ti and Nb-Ti-Zr Alloys at Low Temperatures*, Phys. Stat. Sol. (a) 39, 1977, K67 - K69.

[Fin02] Fink, S.; Ulbricht, A.; Fillunger H.; Bourquard A.; Prevot, M.: *High Voltage Tests of the ITER Toroidal Field Modek Coil Insulation System*, IEEE Transactions on Applied Superconductivity, Vol. 12, Issue 1, 2002, pp. 554 - 557

[Fin04] Fink, S.; Fietz, W. H.; Miri, A. M.; Quan, X.; Ulbricht, A.: *Study of the Transient Votage Behaviour of the Present ITER TF Coil Design for Determination of the Test Voltages and Procedures*, Forschungszentrum Karlsruhe, Scientific Report FZKA 7053, 2004

[Gro62] Grover, F. W.: *Inductance Calculations*, Dover Publications, New York, 1962

[IEC06] IEC 60071-1, *Insulation co-ordination*, International Electrotechnical Commission, Edition 8.0, 2006

[ITER] *www.iter.org*

[IPP09] *http://www.ipp.mpg.de/ippcms/de/for/projekte/w7x/index.html*

[Lib08] Libeyre, P.; Bareyt, B.; Benfatto, I.; Bessette, D.; Gribov, Y.; Mitchell, N.; Sborchia, C.; Simon, F.; Tao, J.: *Minimising Electrical Design Requirements on the ITER Coils*, IEEE Transactions on Applied Superconductivity, vol. 18, (2008), No.2, pp. 479 - 482.

[Max99] *Maxwell 2D Filed Simulator*, Ansoft Corporation, Pittsburgh, 1999

[Mit09] Mitchell, N.: *Scenarios for Coil Power Supply and Cryoplant Analysis*, Release 1.1, ITER_D_2FTVKV, 2009

[Orc07] *OrCAD Capture PSpice 16.0*, Cadence Design Systems Inc, 2007

[PAC1] *Design Description Document 1.1 Magnet, 2 Performance Analysis, 2.1 Conductor Design and Analysis*, N 11 DDD 156 01-07-13 R 0.1

[PAF6A]   *Design Description Document 1.1 Magnet, 2 Performance Analysis, 2.4 Fault and Safety Analysis, Annex 6A, Assessment of the TF coil circuit behaviour during normal and fault conditions*, Bareyt, B.; N 41 RI 34 00-11-01 W 0.1

[PAF6Ba]   *Design Description Document 1.1 Magnet, 2 Performance Analysis, 2.4 Fault and Safety Analysis, Annex 6B, Assessment of the PF coil circuit behaviour during normal and fault conditions*, Bareyt, B.; N 41 RI 35 00-11-01 W 0.1, pp. 2 - 3

[PAF6Bb]   *Design Description Document 1.1 Magnet, 2 Performance Analysis, 2.4 Fault and Safety Analysis, Annex 6B, Assessment of the PF coil circuit behaviour during normal and fault conditions*, Bareyt, B.; N 41 RI 35 00-11-01 W 0.1, Tab. 3.7, p. 13

[PAF6Bc]   *Design Description Document 1.1 Magnet, 2 Performance Analysis, 2.4 Fault and Safety Analysis, Annex 6B, Assessment of the PF coil circuit behaviour during normal and fault conditions*, Bareyt, B.; N 41 RI 35 00-11-01 W 0.1, pp. 6 – 10

[PAF6Bd]   *Design Description Document 1.1 Magnet, 2 Performance Analysis, 2.4 Fault and Safety Analysis, Annex 6B, Assessment of the PF coil circuit behaviour during normal and fault conditions*, Bareyt, B.; N 41 RI 35 00-11-01 W 0.1, pp. 5 - 8.

[PPA3]   *Design Description Document, DDD 1.1 Magnet, 1.3 Procurement Packages, Annex 3: Poloidal Field Coils*

[PPS1]   *Design Description Document, 4.1, Pulsed Power Supplies*, N 41 DDD 16 01-07-06 R 0.3

[PPS1a]   *Design Description Document, 4.1, Pulsed Power Supplies*, N 41 DDD 16 01-07-06 R 0.3, Tab. 1.2.2-6, pp. 75 - 76

[PPS1b]   *Design Description Document, 4.1, Pulsed Power Supplies*, N 41 DDD 16 01-07-06 R 0.3, p. 19

[PPSA1]   *Design Description Document, 4.1 Pulsed Power Supplies, Appendix A, Equipment List and Technical Data Tables*, N 41 DDD 18 01-07-06 R 0.3

[PPSA1a]   *Design Description Document, 4.1 Pulsed Power Supplies, Appendix A, Equipment List and Technical Data Tables*, N 41 DDD 18 01-07-06 R 0.3, p. 23

[PPSA1b]   *Design Description Document, 4.1 Pulsed Power Supplies, Appendix A, Equipment List and Technical Data Tables*, N 41 DDD 18 01-07-06 R 0.3, pp. 30 - 33

[PPSD1] *Pulsed Power Supply Drawings*, 41.0024.0001, 41.0025.0001, 41.0026.0001, 41.0027.0001, 41.0028.0001, 41.0029.0001, 41.0030.0001, 41.0031.0001

[PPSD2] *Pulsed Power Supply Drawings*, 41.0040.0001

[Rim00] Rimikis, A.; Hornung, F.; Schneider, Th.: *High Field Magnet Facilities and Projects at the Forschungszentrum Karlruhe,* IEEE Transactions on Applied Superconductivity, vol. 10, (2000), No.1, pp. 1542 - 1544.

[SLBa] Design Description Document, DDD 6.2 Site Layout and Buildings – Introduction, N 62 DDD 22 01-07-18 R 0.1

[SLBD1] *Drawings DDD 6. 2 Site Layout and Buildings*, 62.377.0001, 62.0333.0001-0015, 62.0357.0001-0004

[Ulb05] Ulbricht, A.; et al.: *The ITER toroidal field model coil project*, Fusion Engineering and Design 73, (2005), pp. 189 - 327

[West07] Westcode, Westpack Phase Control Thyristor Type N1174JK200 to N1174JK220, Data sheet, February 2007

[Win09] Winkler, A.; Fietz, W. H.; Fink, S.; Noe, M.: *Calculation of Transient Electrical Behaviour of ITER PF Coils*, Fusion Engineering and Design 84, (2009), pp. 1979 – 1981.